CAMBODIA'S CONTESTED FOREST DOMAIN

ASIAN STUDIES
Caroline S. Hau, *Series Editor*

{ CAMBODIA'S CONTESTED FOREST DOMAIN }
THE ROLE OF COMMUNITY FORESTRY
IN THE NEW MILLENNIUM

{ MARK POFFENBERGER, EDITOR }

ATENEO DE MANILA UNIVERSITY PRESS

ATENEO DE MANILA UNIVERSITY PRESS
Bellarmine Hall, Katipunan Avenue
Loyola Heights, Quezon City
P.O. Box 154, 1099 Manila, Philippines
Tel.: (632) 426-59-84 / Fax (632) 426-59-09
E-mail: unipress@admu.edu.ph
Website: www.ateneopress.org

Book design by Sam Immanuel R. Macaisa
Cover design by Karl Fredrick M. Castro

The publication assistance of Community Forestry International (CFI) is
gratefully acknowledged.

The National Library of the Philippines CIP Data

Recommended entry:

    Cambodia's contested forest domain : the role of
      community forestry in the new millennium /
      Mark Poffenberger, editor. -- Quezon City :
      Ateneo de Manila University Press, 2013.
        p. ; cm.

    ISBN 978-971-550-653-3

      1. Community forestry--Cambodia. 2. Forest
management--Citizen participation--Cambodia.
3. Forest policy--Cambodia. I. Poffenberger,
Mark, editor.

SD657.C35 333.7509596 2013 P220120808

CONTENTS

PART I: THE CAMBODIAN CONTEXT

PART II: COMMUNITY FORESTRY IN THE FIELD

PART III: THE VALUE OF FOREST MANAGEMENT TRADITIONS

APPENDIX: SPECIAL SECTION ON FOREST LAWS, POLICIES, AND PROGRAMS

LIST OF TABLES

LIST OF FIGURES

LIST OF BOXES

ACRONYMS

ABE	Association of Buddhists for the Environment
ADB	Asian Development Bank
AFSC	American Friends Service Committee
ANR	Assisted Natural Regeneration
ARC	Alliance of Religions and Conservation
ASEAN	Association of Southeast Asian Nations
CBFM	Community-based Forest Management
CBNRM	Community-based Natural Resource Management
CBSFLMP	Capacity Building for Sustainable Forest and Land Management Project
CCA	Community Conservation Area
CCBA	Climate and Community and Biodiversity Alliance
CDA	Children's Development Association
CDM	Clean Development Mechanism
CF	Community Forestry
CFA	Community Forest Agreement
CFAC	Community Forestry Alliance for Cambodia
CFI	Community Forestry International
CFMA	Community Forestry Management Agreement
CFMC	Community Forestry Management Committee
CFMP	Community Forestry Management Plan
CFSD	Community Forestry Sub-Decree
CFU	Community Forestry Unit
CFUG	Community Forestry User Group
CLUP	Commune Land Use Plan
CMDGs	Cambodia Millennium Development Goals
CNPA	Community Nature Protected Area
COM	Council of Ministers
CPA	Community Protected Area
CPP	Cambodian People's Party
CTIA	Cambodia Timber Industry Association
DAI	Development Alternatives International
DF&W	Department of Forests and Wildlife (now the Forestry Administration)
ELC	Economic Land Concession
FA	Forestry Administration
FACT	Fisheries Action Coalition Team
FAO	Food and Agricultural Organization
FCMCPP	Forest Concession Management and Control Pilot Project
FCMRP	Forest Crimes Monitoring and Reporting Project
FACT	Fisheries Action Coalition Team
GDANCP	General Department of Administration for Nature Conservation and Protection

GHG	Greenhouse Gases
GIS	Geographic Information System
GPS	Global Positioning System
GTZ	German Technical Assistance
ICSO	Indigenous Community Support Organization
JSDF	Japanese Social Development Fund
LEK	Local Ecological Knowledge
LIS	Land Information System
MAFF	Ministry of Agriculture, Forestry and Fisheries
MCF	Monks Community Forest
MES	Markets for Ecosystem Services
MLMUPC	Ministry of Land Management, Urban Planning, and Construction
MOE	Ministry of Environment
MOJ	Ministry of Justice
NCFP	National Community Forestry Program
NFP	National Forest Program
NPA	National Protected Areas
NPASMP	National Protected Areas Strategic Management Plan
NRM	Natural Resource Management
NSDP	National Strategic Development Plan
NTFP	Non-Timber Forest Products
PA	Protected Area
PES	Payment for Environmental/Ecosystem Services
PFE	Permanent Forest Estate
PFR	Permanent Forest Reserve
PLUP	Participatory Land Use Planning
PRA	Participatory Rural Appraisal
RECOFTC	Regional Community Forestry Training Center
REDD	Reducing Emissions from Deforestation and Forest Degradation
RGC	Royal Government of Cambodia
SFM	Sustainable Forest Management
SFMP	Sustainable Forest Management Plan
SGP	Seed Grants Program
SLC	Social Land Concession
TGC	Terra Global Capital
TSEMP	Tonle Sap Environmental Management Project
TWG-F&E	Technical Working Group on Forests and the Environment
UNFCCC	United Nations Framework Convention on Climate Change
UNTAC	United Nations Transitional Authority in Cambodia
VCS	Voluntary Carbon Standard
VNP	Virachey National Park
WWF	World Wide Fund for Nature

ACKNOWLEDGMENTS

THIS BOOK IS DEDICATED TO CAMBODIA'S FOREST COMMUNITIES in recognition of the important role they are playing in managing the nation's forests. For centuries they have acted as the stewards of the region's rich forest ecosystems. The last century has challenged their capacity to protect and manage forestlands, yet many villages have persisted to ensure that these forests remain intact. With the dawning of the twenty-first century, they are now positioned to have their forest rights legally recognized. These grassroots community forest management initiatives have both inspired and guided the writing of this book.

Special recognition is given for the pioneering community forestry work of Cambodia's Forestry Administration, including H.E. Ty Sokhun, Chea Sam Ang, Ly Chou Beang, Sokh Heng, Lao Sethaphal, Prak Marina, Long Ratanakoma, Min Bunnara, Sy Ramony, Nup Sothea, Dy Sophy, and many other individual foresters working in the field. Much of the early efforts and continuing innovations have been contributed by Cambodians working with other government agencies, as well as international, national, and local NGOs, including Ken Serey Rotha, Seng Teak, Srey Marona, Hou Kalyan, Sao Vansey, Sok Khim, Lim Solinn, Yeng Virak, Chean Thayth, Yang Phirom, Ou Sopheary, and many others.

Over the past twenty years, Cambodia's community forestry program has also been strengthened through the efforts of individuals from other countries, including Patrick Evans, Toby Carson, David Ashwell, Kiran Paudyal, Graeme Brown, Amanda Bradley, Russell Peterson, Robert Oberndorf, Lars Lund, Vaneska Litz, Yukiko Yonekura, Joe Walston, Goseki Kazuhiro, Doug Henderson, Ashish John, Jurgen Fictenau, Tom Evans, Andrew Maxwell, and numerous other individuals and organizations.

Community Forestry International (CFI) would like to thank the John D. and Catherine T. MacArthur Foundations and the United

States Agency for International Development for supporting CFI's work in Cambodia over the years. We would also like to express our appreciation to our partner organizations that have supported community forestry field projects and research including Oxfam G.B., RECOFTC, WWF, CI, PACT, WCS, FFI, GTZ, Project Concern, MCC, CDRI, and many others. Special acknowledgement is also important for the collaborative research programs undertaken by the Community-based Natural Resources Management Learning Institute (CBNRM-LI) and the Cambodian Development and Research Institute (CDRI).

The editor would like to thank the chapter authors who made this book possible, with special thanks to Ian Baird, Chantal Elkin, Jeff Fox, Long Ratanakoma, Melissa Marschke, Robert Oberndorf, Kathryn Smith-Hanssen, Abu Hassan Ahmad, Zainal Ariffin Ahmad, and Edward Maningo. Thanks are also due to H.E. Ty Sokhun who provided important insights and inputs into its development. We are also grateful to the editorial staff at the Ateneo de Manila University Press for organizing the reviews of the book and its publication, with special thanks to Maricor Baytion and Tina Castro. Finally, we want to recognize the important contribution of Kathryn Smith-Hanssen who worked with the editor to guide the development of the manuscript over the past two years, providing thoughtful comments and reviews, coordinating with authors, as well as careful editing.

In closing, it is important to acknowledge the tremendous effort being made by hundreds of rural communities across Cambodia that contribute to forest protection, conservation, and the sustainable management of the planet's natural ecosystems. They require the support of national governments and the international community.

—Mark Poffenberger, Editor

INTRODUCTION

For millennia, abundant rainfall has helped create Cambodia's lush vegetation and biodiversity. Forests were inhabited by small groups of hunters and gatherers, as well as isolated farming communities. Each cultural community developed unique sets of knowledge regarding the local ecology and the uses of hundreds of plants and animals, as well as ways to manipulate the forest to optimize its productivity. Indigenous management systems and the local knowledge that supports them have been passed down for generations. However, the Cambodian kingdoms began to extend control over forests within their periphery as early as the 7th century CE. Stone inscriptions from 639 CE during the reign of King Per Vearmon II indicate there were forestry professionals in charge of each region. Royal tax collection codes relating to forest exploitation were functioning during the time of King Ang Duong (1845–1859).[1]

In 1863, Cambodia became a "protectorate" of France, after which the French colonial officials began controlling forest policies.[2] By 1907, Mr. Du Camp, Director of Water and Forest Conservation for the Ministry of Forest Indochina, declared forest policy goals that included banning shifting cultivation and "illegal logging;" however, those policies could not be widely implemented due to the lack of management and equipment.[3] Nonetheless, the French colonial government did establish a national forestry agency, law and policy framework, and technical orientation that continue to influence contemporary national strategies for natural resource management. Perhaps the most significant influence is the concept of "state forest land."

Nationalization of forest lands during the colonial period was a common practice throughout South and Southeast Asia. These policies were first implemented by the Spanish in the Philippines in the late 16th century, followed by the Dutch in Indonesia, the British in India and

Burma, and the French in Indochina. The forest tenure rights of indigenous communities and local rulers were often ignored in the process, creating a fundamental cause of "contested domain" that remains to the present in many parts of the region. Although local rulers and kingdoms may have controlled forests in areas near their capitals, much of Cambodia and other outlying forest regions in Southeast Asia were beyond the reach of the court and largely under the control of forest communities.[4] The colonial forest paradigm was therefore new, not only in claiming state ownership of virtually all forests, but introducing new bureaucratic and technocratic systems for the "scientific management" of forests. Equally influential was the concept that forest lands could be utilized and viewed as a commodity for exploitation and revenue generation, which contrasted sharply with traditional views of forests as resources held in trust for future generations.

From the mid-19th century, forest-dependent communities in Cambodia and elsewhere in Southeast Asia were often stigmatized by government as destroyers of the forest, "encroachers" on public lands, especially indigenous communities, who were often ethnic minorities. Traditional farming systems, such as shifting cultivation, were often condemned and banned. Yet, while the laws and policies of state control have been established for over a century, it is their implementation in recent decades that has both accelerated forest loss and created conflict over resource control. As David Korten notes throughout Asia, "Public development efforts of the past few decades have seen increasing extension of state authority throughout Asia into affairs once the preserve of local custom and control."[5] Decisions by government agencies to allocate "state forest lands" for commercial logging and economic concessions rarely involve substantive discussions with local communities, planting the seeds of social conflict which remain up to the present time.

Over the past several hundred years in Cambodia, these human-forest relationships have changed at an accelerating pace, especially in recent decades. Nationalization of forestland during the colonial period, followed by periods of political instability, expanding populations, the demands of global markets, changing technologies, and ever-improved access, have opened the nation's forests and driven rapid exploitation. This has resulted in forest cover declining from over 90 percent in the pre-colonial era, to 73 percent in 1965, and to approximately 57 percent today. Further, the quality of the forest has deteriorated due to illegal logging, forest fires, and other disturbances disrupting a range of environmental services, including hydrological functions, critical biodiversity habitat, and non-timber forest product (NTFP) productivity essential for rural community livelihoods.

Since the colonial era, the vast majority of the region's forestland has been, at least nominally, under the legal stewardship of national forestry agencies. The primary mandate for forest management has been the generation of revenues for the state, primarily through the leasing of state public forest lands to private enterprises and charging royalties on timber. By the late 1990s, over 60 percent of Cambodia's forestlands were leased to industrial logging concessionaires. Yet, Cambodia, like a number of other ASEAN nations, has been disappointed by the failure of commercial logging enterprises to perform sustainably and to contribute meaningfully to the national and local economy. In addition to degradation inflicted by unsustainable logging practices, other forestlands have been cleared for agricultural use, either by large concessionaires, migrant farmers, or local residents. While some forest conversion has been an inevitable result of population growth and development, large areas have been unsustainably exploited and left in degraded condition, and they now provide few benefits either in terms of hydrology, environmental services, or economic productivity.

As the State extended its control over natural resources, it overestimated its own capacity to manage forests, while underestimating the ability of traditional systems to effectively manage local forests to meet their own needs.[6] The trend to establish state control over resources is driven by a process of modernization and state building reflecting the larger global paradigm,[7] the national government's need to generate revenues for economic growth, as well as aspects of the Cambodia's political economy that has allowed some elite families to personally profit from forest resources. However, this trend for centralized state control over forests has begun to shift over the past decade as an alternative form of participatory forest management have been adopted as government policy. It is increasingly apparent that if national macro and micro policies are formulated that can achieve the multiple objectives of sustainable resource management, economic growth, and social equity, while reducing conflicts, some compromises are required. As Korten comments:

In the face of rapidly increasing pressures on a finite resource base generated by growing populations and rising aspirations, there is a need for substantial and rapid evolution of existing resource management systems to support sustainable intensification of resource use. It is unlikely that traditional village communities can accomplish this rapid evolutionary change on their own. But neither can the state accomplish it entirely through its bureaucratic instrumentalities. There must evolve a more dynamic partnership arrangement building from the existing capacities and evident self-interest of the local community and complemented by the ability of the state to support the development of enabling policies and institutional linkages.[8]

A fundamental forest paradigm shift is taking place in Cambodia and some other Southeast Asian nations, transforming the ways in which governments perceive and manage their forest environments. For over a century, the dominant model has dictated that professional foresters should hold primary control over the majority of each nation's land area, typically ranging from 30 to 70 percent, under the ultimate direction of the central government, with utilization rights usually leased to private companies. The failure of these operations to sustainably manage their concession areas resulted in rapid deforestation and degradation, together with social marginalization, creating an international outcry that led to the cancellation or suspension of virtually all Cambodia's major concessions by 2000. The resulting management vacuum created by the closure of six million hectares of timber leases has allowed for new paradigms to emerge, including the national community forestry and conservation movement, as well as a diverse array of conservation initiatives.

The Cambodian constitution notes the rights of communities to own land and forests privately and collectively, and to form associations. This was further supported through the approval of the National Community Forestry Program (NCFP) and the National Forest Program (NFP), which together provide a strong policy framework that supports CF legislation. Collectively, these actions reflect a broad-based shift away from natural resource management (NRM) laws, policies, and management systems, many established during the colonial period and relied upon during early independence, towards twenty-first-century strategies to empower resource-dependent communities.

These strategies, however, are competing with private sector investments in the form of economic concessions that target state public forestlands—investments that many senior government policymakers view as essential for rapid economic growth. Cambodian planners are evaluating proposed economic land concessions (ELCs) for sugar and rubber plantations, and mining in relation to the value of retaining forests under protected areas, community forest management, and other management systems geared toward national conservation goals as well as the needs of rural communities.

Strategies to conserve Cambodia's unique lowland and montane forest ecosystems are challenged by an accelerating reallocation of public forestlands, as well as an overheated property market dominated by political and private sector elites. With the demarcation of the permanent forest estate still in an early phase, forestry and environmental agencies affiliated with government, communities, and NGOs are confronted by stakeholders who wish to utilize forest environments for commercial purposes. It is apparent that Cambodia's forests are "contested domain," as the struggle for control plays out in one of mainland Southeast Asia's last frontier forest regions.

As these competing forces strive to gain rights over the country's forest lands, it is clear that communities are at a disadvantage in terms of their economic and political leverage. By mid-2012, one assessment found that economic land concessions controlled by agri-industrial companies held more than 2 million hectares of state land with another 1.9 million hectares allocated to mining companies in exploration rights, representing a total of 22 percent of Cambodia's total land area.[9] By contrast, in 2011 the National Community Forestry Program (NCFP) recognized 441 CF groups that managed 389,000 hectares representing 4% of the forest area. While the NCFP has started recently, the national target for CF coverage is 2 million hectares and a 2002 assessment estimated that up to 8.4 million hectares of forest could potentially be brought under community management.[10] The competing paradigms for rights to manage state forest lands appear to favor economic land concessions which cover larger areas and can be processed more quickly than CF agreements. When concessions enter into forest areas to initiate land clearing, conflicts arise. Since 2000, according to the NGO, ADHOC, 700,000 people nationwide have been involved in land disputes after private companies gained control of their local forests and farmlands.[11] On May 8, 2012 Cambodian Prime Minister Hun Sen announced that he would temporarily suspend new economic land concessions and that he would revoke concessions involved in land grabbing and illegal logging.[12] This government action took place soon after a prominent NGO leader was shot and national media coverage of rapid concession issuance grew. These events graphically illustrate the nature of Cambodia's Contested Forest Domain and the vulnerable position and potential role community forestry can play in establishing more equitable, sustainable systems of natural resource management.

As this book indicates, Cambodia faces a diverse set of challenges as it explores ways to link the formal forestry sector to informal and emerging initiatives of rural communities. This is a complex task, as the forestry sector in Southeast Asia, as in many parts of the world, continues to be characterized by a technocratic, hierarchical, and bureaucratic orientation to management, and by a need to retain control and authority. The informal forest cultures of the region, by contrast, emphasize a highly decentralized, communal orientation based on a multiplicity of uses.

The authors of this book take the position that one of the most promising approaches to restoring and conserving Cambodia's forests and encouraging sustainable and equitable economic growth will be through the engagement of rural communities in their management. This strategy does not represent a "magic bullet" that can quickly halt or reverse forest loss, nor does it preclude opportunities for commercial enterprises on forest lands or for strict conservation. It is a process that provides a pathway to formalize the role of forest communities as natural resource managers in partnership with local and national government, while focusing support to improve the productivity management systems used to address rural poverty. Equally important, community forestry (CF) provides one approach to the demarcation, classification, and registration of the permanent forest estate (PFE). This important process is a key element in clarifying land and resource tenure and resolving social conflict.

This book presents a variety of perspectives on Cambodia's ongoing forest transition and the role communities, government, and civil society are playing in the process. Part I provides an overview of the recent history of forest management in Cambodia, emphasizing important changes in the forest sector and the emergence of a new national policy and legal environment. Chapter 1 examines forest trends over the past twenty years as national planners and policymakers attempt to rethink the country's forest management strategy. Experiences with the commercial timber concession program of the 1990s are reviewed, as well as the social and environmental impact of this strategy. The chapter proceeds to describe the events leading up to the moratorium on logging concessions and the subsequent emergence of a range of management options, including community forestry. The author identifies a variety of emerging policies and programs being pursued by the Royal Government of Cambodia (RGC), some of which are designed to maximize economic growth through commercial concessions, while others support the decentralization of the forest estate vesting rural communities with management rights to forests, fisheries, and state land.

Part II presents diverse examples of communities and forest management and how they are responding to a changing world. Beginning with a national overview of some aspects of the informal and formal community forestry movement, this section proceeds to examine experiences from the northwest, the Tonle Sap Basin, and northeast Cambodia. Chapter 2 reviews assessments of rural dependence on Cambodia's forest resources. The review indicates that extensive areas are under community use with varying degrees of defacto community control. The authors proceed to discuss the evolution of community forestry support programs developed by NGOs and by the RGC and how these initiatives have led to the establishment of a national program with formalized CF groups functioning with agreements from the Forestry Administration. The chapter concludes with a discussion of the economic importance of forests for resident communities and how community forestry could help resolve resource conflicts, promote economic development, and conserve Cambodia's forest ecosystems.

Chapter 3 focuses on early learning from Cambodia's first project to reduce emissions from deforestation and forest degradation, or REDD. The emergence of international voluntary markets for forest carbon has created a potential long-term mechanism for financing sustainable forest management. In the case of Cambodia, where recent deforestation rates have ranged from 0.5 to 2 percent annually, there is substantial opportunity to slow forest loss and generate carbon credits by modifying forestland allocation policies, as well as making field-level management improvements. The author notes that the project is an important "proof of concept" for REDD more generally, as well as a feasibility test to determine if PES project models may be a viable way to finance the NCFP and the national forest sector transition. The chapter describes the project design process, drivers of deforestation, and mitigation strategies, as well as problems encountered along the way. Special attention is given to problems in securing financing for early project development, questions about benefit sharing, and constraints faced in working with international certification processes and carbon markets.

Chapter 4 explores changing management systems for Cambodia's unique flood forests and fisheries by examining the experiences of Kompong Phluk village located in the flood forests of Tonle Sap Lake. Flood forests surround and protect Cambodia's Great Lake, the largest freshwater lake in Southeast Asia and one of the world's most productive inland fisheries. In 1998, over 1.2 million people lived within the floodplain. In response to tensions and conflicts over access to the fishery, over the past twenty years the Department of Fisheries has developed new policies and programs to devolve management of fishing lots and flood forests to resident villages. This chapter describes key policy actions and how they have impacted Kompong Phluk, a flood forest community in Siem Reap Province, that has pioneered community-based fisheries and flood forest management. As the author notes, while fishing communities have successfully protected and restored forests and gained greater resource security, they face new challenges from climate change, falling fish productivity, and competition from economic development projects and other user groups.

Chapter 5 assesses how socioeconomic forces have been impacting forest-dependent communities in northeast Cambodia over the past twenty years, as population expansion, in-migration, land speculation, and market penetration rapidly change the ways in which indigenous communities manage their forest and agricultural lands. The study of three traditional communities finds that the growth of land markets and the entrance of land speculators are eroding traditional systems of communal forestry and land tenure, while the introduction of cash crops is transforming land use systems. While two of the villages have experienced rapid and uncontrolled deforestation and land alienation, one of the three indigenous communities is sustainably managing their traditional lands and forests. The chapter concludes that support for the capacity development of village-based institutions, community land use planning, and communal tenure strengthening can help stabilize forest cover, thus reducing the rate of land alienation and allowing for sustainable transitions in land- and forest-based production systems.

Part III explores the value of forest management traditions in Cambodia, emphasizing both the long history and depth of local knowledge, as well as the cultural values that support conservation and sustainable use. Cambodia's forest history over the past 150 years has been dynamic and traumatic. Located in the heart of mainland Southeast Asia's tropical hill, plain, and flood forests, Cambodia possesses diverse forest ecosystems, immense biodiversity, and varied traditional systems of resource stewardship. Indigenous Khmer community and hill tribe management practices continue, though they have been overshadowed by a "scientific forestry" paradigm introduced over a century ago.

At the same time, economic, political, and demographic forces are reshaping the nation's landscape, placing immense pressure on the natural environment. However, in the midst of this complex, dynamic, and damaging process, the country is moving towards new forms of more sustainable management centered in rural communities. This section explores both traditional forms of forest stewardships and how local forest management systems are responding to change, including the emergence of civil society initiatives such as the Buddhist environmental movements.

Chapter 6 documents the experiences of a group of Buddhist monks who are pioneering the establishment of one of Cambodia's largest community forests, an 18,000-hectare tract in Oddar Meanchey Province. Unlike other NGOs and civil society initiatives supporting community forestry, the Monk's Conservation Forest (MCF) represents a Khmer approach to forest protection based on indigenous environmental values and beliefs, as well as newer Buddhist environmental activism also occurring in Thailand. With modest external support, and under the leadership of a charismatic young monk, the monks have mobilized their monastery (pagoda) and local villages to defend the forest against encroachment by illegal loggers, soldiers, and land speculators. This chapter examines the activities undertaken, the challenges this grassroots initiative faces, as well as the broader Buddhist principles upon which the forest protection movement is based. The author concludes that Buddhist environmentalism represents an approach that could be widely replicated in Cambodia and that is consistent with Khmer values and traditions.

Chapter 7 examines the complex nature of the indigenous Kavet system of land and forest classification and knowledge. The author illustrates how an understanding of ethnoecology can be helpful to land use planning processes in the context of protected area management. In this case study, the Kavet live in and around Virachey National Park (VNP) and have been engaged in swidden agriculture for generations. The author suggests that the relatively sustainable shifting farming systems developed by the Kavet for bamboo forests may be a viable land use practice within the VNP. The chapter concludes that local ecological knowledge (LEK) should be integrated into protected area management. This will require government managers to make a greater effort to understand indigenous land use classification systems and practices, and to engage resident people in management decision-making.

The Conclusion provides a brief summary of Cambodia's forest sector transition, drawing on each chapter. The conclusion notes that participatory resource management has the potential to address many of the problems confronting Cambodia's forests and forest-dependent people, including social conflicts, illegal logging, conversion to agriculture, and forestland speculations. Resident communities often possess both the capacity to sustainably manage these valuable resources, as well as the values to conserve them for future generations. The summary section stresses the need for government, development agencies, and civil society groups to cooperate in order to make the NCFP and other people-oriented initiatives a success.

The book ends with a Special Section on Forest Laws, Policies and Programs. This part provides readers with additional perspectives and in-depth information regarding RGC's strategy to support community forestry. This special section includes an overview of the national forestry law and policy actions by Rob Oberndorf, JD, an environmental lawyer who provided technical assistance to the RGC and has written widely on the subject. This section examines the legal frameworks and policy actions that are enabling the devolution of resource management rights in Cambodia. It documents important legislation and policies that were approved during the past ten years, analyzing the significance of the changing directions in law and policy and their potential impact on national natural resource management systems. The rights of communities to derive subsistence goods from state lands and aquatic resources were articulated in the 1993 constitution, and in subsequent forest and land laws. While economic growth objectives resulted in a heavy emphasis on the commercial exploitation of natural resources in Cambodia during the 1990s, there has been a clear shift towards greater community involvement in resource management over the past decade. As a whole, a growing body of Cambodia's laws, royal decrees, sub-decrees, and implementing orders provide an impressive body of legal decisions that support the rights of local governments and communities to engage in natural resource management.

The second section was written by Abu Hassan Ahmad, Zainal Ariffin Ahmad, and Edward Maningo. This section presents a broad overview of new directions in national forest management goals and policies. The authors describe the importance of establishing sustainable management systems that address the needs of Cambodia's large and growing rural population. Noting that rural Cambodians remain heavily dependent on natural resources, policies and programs are required that extend secure forest tenure rights, allowing rural communities to invest in developing these important national assets. The authors explain that past community forestry initiatives tended to emphasize reforestation and forest protection, but new policy goals stress the need to make forests responsive to the needs of rural communities.

The third section by Long Ratanakoma and Edwin Payuan discusses the implementation of the NCFP, including the eleven steps that lead to CF agreements and full implementation. This section also discusses the impact and coverage of the NCFP to date. The authors note that those regions of Cambodia that received substantial external support for CF development, particularly the North Tonle Sap Inspectorate (both under the earlier FAO project [UN Food and Agricultural Organization] and the current Oddar Meanchey REDD project [Reducing Emissions from Deforestation and Forest Degration]), have experienced accelerated expansion and formalization of their CF groups. Both projects involved partnerships engaging the communities, The Forestry Administration (FA), local NGOs, and international organizations.

Made up of different points of view on the success and challenges of community forestry, ultimately this book seeks to summarize the changes taking place in Cambodia's forest sector over the past twenty years, emphasizing the important roles rural communities have played in forest management in the past, present, and future. The book was developed to better inform a general audience interested in this topic, with reference to students and researchers, development workers, foresters, and government planners and policy makers.

The book raises a number of important issues and questions regarding how the RGC will move forward in balancing programs like the NCFP, which support the devolution and empowerment of communities to manage state forest lands, with the differing development approaches such as those reflected in allocating state forest lands for economic land concessions and mining exploration rights. As a number of authors note, these decisions not only reflect natural resource management goals and economic development considerations, but are increasingly of a political nature. Recent public demonstrations and mass media coverage regarding social conflicts between rural communities, private sector firms, and government dramatically illustrate this broader national dialogue regarding the future of Cambodia's forests.[13]

NOTES

1. Hung Sun Tra and Arvid Sloth "Review of International Forest Policy Development and Cambodia's Role, Involvement and Potential Benefit" (Phnom Penh: Forestry Administration, April 2007).

2. Milton Osborne, *The Mekong* (New York: Grove Press, 2000), 120.

3. Hun Sun Tra, "Review of International Forest Policy Development," 5.

4. Mark Poffenberger, *Keepers of the Forest* (New Haven, Conn.: Kumarian Press, 1990), 11.

5. David Korten, *Community Management: Asian Experiences and Perspectives* (New Haven, Conn: Kumarian Press, 1987), 1.

6. Ibid.

7. Ibid.

8. Ibid.

9. Paul Vrieze and Kuch Naren, "Carving Up Cambodia One Concession at a Time," *The Cambodia Daily* (March 10–11, 2012).

10. Jurgen Fichtenau, Ly Chou Beang, Nup Sothea, and Dy Sophy, "Assessment of Ongoing Community Forestry Initiatives in Camobdia" (Phnom Penh: DFoW//GTZ, 2003).

11. Vrieze and Naren, "Carving Up Cambodia," 7.

12. www.rainforests.mongabay.com/20cambodia.htm (May 25, 2012).

13. Vrieze and Naren, "Carving Up," 7.

PART I

THE CAMBODIAN CONTEXT

CAMBODIA'S CONTESTED FOREST DOMAIN: A HISTORICAL PERSPECTIVE

MARK POFFENBERGER

OVER THE PAST CENTURY, CAMBODIA HAS LOST 30 PERCENT OF its forest cover, while much of its primary forest has been degraded, losing biomass, biodiversity, economic value, and key hydrological and climatic functions. Cambodia's experience is not unique, as throughout mainland Southeast Asia, primary forest is being degraded or deforested at a rapid rate. In the Lower Mekong sub-region, it is estimated that tens of millions of hectares of forests were significantly modified or removed during the 1980s and 1990s.[1] While commercial logging is often the initial cause of declining primary forest cover, rising demand for perennial cash crops like coffee, palm oil, and rubber increases the pressure to convert remaining forests to commercial cash crops. In Cambodia, and throughout the region, logged-over sites are frequently occupied by migrant homesteaders and the commercialization of land markets has brought speculators who have further accelerated these drivers of deforestation and forest degradation. Economic growth and development policies have largely supported the transfer of state forest lands to private firms through the issuance of economic land concessions. As Mittelman observes:

Policies encourage such conversion, and longer-term land tenure security may be available through forest conversion to agricultural tree plantation. Such areas are favored targets for urban speculation.[2]

Community forestry represents a strategy to stabilize Cambodia's natural forests while ensuring that rural villages, heavily dependent on forest resources for their livelihoods, will not see their land alienated and the landscape transformed. It has been estimated that since 1973, eleven million hectares of Cambodia's forests had been degraded through illegal logging, charcoal making, fuelwood collection, and other disturbances. By 1993, 54 percent of Cambodia's land area was classified as medium-low density cover.[3] While a substantial proportion of this area is converted to agriculture and non-forest uses each year, in many areas where community forestry groups are active, forests are being protected and are regenerating.

While the forests of the Lower Mekong sub-region are disappearing rapidly, including those in Cambodia, there is hope that community forestry can stabilize the situation in an environmentally responsive and socially equitable and just manner. Some analysts note that recently revised forest policies, such as the new Cambodian Land Law and Forest Law, delegate unprecedented authority to communities for managing forestlands. But it remains to be seen whether government leaders will demonstrate the political will to implement them.

A challenge to successful implementation of new pro-community forestry laws and policies is their integration into national development plans and donor priorities. According to the Rectangular Strategy for Growth, Employment, Equity, and Efficiency, which was launched in mid-2004 by the Royal Government of Cambodia (RGC) as the national development agenda, government planners envision promoting economic growth through agriculture, infrastructure, and private sector expansion. With limited financial resources, the RGC is dependent on the private sector and development assistance to finance many aspects of this strategy. In other countries in the region, this approach has led to a commoditization of agriculture, the expansion of speculative land markets, and a clearing of natural forests throughout the Mekong region, often displacing forest-dependent peoples and their land use systems.

Development plans and policies that further support private sector investors, who are already rapidly acquiring land and forest control, often illegally, will likely accelerate deforestation and land alienation. An important step in this process is enhancing the land and forest tenure security of the local population. While the necessary legal instruments have already been ratified under the national Land and Forest Laws, it is still necessary to establish a strong political will for implementing the policy and organizing the financial resources required to proceed with their implementation.

Forest policies provide a course of action adopted by a government to achieve specific environmental, economic, and social goals. Cambodia has experienced significant shifts in its forest policies over the past twenty years. This chapter describes how the forest sector has changed over the past two decades as commercial timber concessions boomed, and then decreased when concessions failed to protect the environment, avoid social conflict, and generate revenues for the state. These events and policy changes provided new opportunities for the emergence of the National Community Forestry Program (NCFP).

In the post–World War II era, the forest sector was viewed in many developing countries in Southeast Asia as an important component supporting early economic growth. Emerging national governments with limited revenue-generating capacity identified the leasing of public forestlands to public and private logging concessions as a means to finance government administration and the military. The RGC military extracted significant revenues from the forestry sector from log harvesting, transport, processing, and trade.[4] Between 1994 and 1997, the RGC leased over thirty forest concessions covering an area of 6.5 million hectares, and representing almost 60 percent of the nation's forests. By leasing state forests to commercial timber companies, the political leadership hoped to delegate responsibility for their management and generate revenue for national development, both of which were urgently needed after the widespread social and economic disruption created by the civil war.[5] During the same period, the RGC moved to create and expand protected areas throughout the country. Both the production and protection of forest systems, however, suffered from a dearth of technical and financial resources.

By 1996, concerns were raised about the sustainability of industrial logging, as well as the viability of protected areas. A World Bank–funded Forest Policy Assessment report published in April 1996 noted that Cambodia had experienced deforestation of one million hectares and degradation of three million hectares of forests over the past thirty years (1965–1995), reflecting a loss or deterioration of about 35 percent of its forest ecosystem.[6] By the mid-1990s, bilateral and multilateral agencies supporting Cambodia's forest sector were concerned about the forest loss and discouraged by high levels of illegal felling, yet held firm in their belief that sustainable commercial timber concessions could be developed.

As a consequence, the World Bank, FAO, and the United Nations Development Programme (UNDP), as well as the Department for International Development (DFID) and the Danish International Development Assistance (DANIDA), provided funding for a range of forest policy reform projects, forest sector reviews, and financing for sustainable forest management initiatives. At the same time, the RGC took many policy actions to control commercial logging and, later, to establish alternative management systems. In this way, the national government, development agencies, International NGOs (INGOs), NGOs, and civil society reshaped Cambodia's forest sector, a process that keeps reinventing itself as new opportunities and technologies become available.

FOREST CONSERVATION

Cambodia established a protected area (PA) system during the French colonial period. Prior to 1957, one-third of the country had been set aside in 173 forest reserves and six wildlife areas.[7] The six wildlife reserves covered approximately 2.2 million hectares, with an additional 10,800 hectares set aside around the temples of Angkor, which became Southeast Asia's first national park in 1925. While there were plans to create a national park system based on the wildlife reserves as early as 1966, the civil war delayed this initiative, and the protected areas received little attention until the early 1990s.

With the signing of the Paris Peace Accord in 1991, the RGC was able to begin addressing the need for forest conservation. In 1993, a royal decree was issued by the king to create and designate twenty-three protected areas under the authority of the Ministry of Environment (MOE). By 1998, Cambodia had set aside nearly 20 percent (3,568,100 ha) of its land area under protected area management and, as a proportion of national forest, had established one of the most extensive protected area management systems in Asia.[8] The MOE controlled much of this area, which was managed as seven national parks, ten wildlife sanctuaries, three protected landscapes, and three multiple use management areas. However, the MOE had limited staff, funds, and resources to prevent encroachment by commercial timber concessions and illegal loggers.

The Department of Fisheries was responsible for the flood forests around Tonle Sap (the Great Lake), as well as for coastal mangrove forests that total around one million hectares. The Department of Forests and Wildlife (DF&W), now the Forestry Administration (FA), controlled 60 percent of the land area of Cambodia, a substantial proportion designated as protected forest. Yet, DF&W also struggled with a lack of funding. In 1997, for example, the RGC only provided US$489,746 for the entire department, representing approximately 17 percent of the agency's requested budget that year. As a result, the department's 1,622 staff, 55 percent of which were located in the provinces, had almost no funding for salaries and implementing field management activities.[9] Of twenty-four vehicles, only twelve were operational.

While both the MOE and the FA lack sufficient financing to carry out conservation activities, pressures on the protected area system are substantial. Illegal logging within protected areas and encroachment by concessions have been an ongoing problem. Global Witness monitored Cambodia's national parks, wildlife sanctuaries, and protected forests from 2001 to 2002 and found illegal logging in virtually all protected areas. In 2001, in Phnom Samkos Wildlife Sanctuary, nineteen Thai loggers were arrested, while sixteen illegal sawmills were found operating in the Aural Wildlife Sanctuary. The report noted that newly hired rangers lacked training, as well as the political and financial support and basic equipment required to curb illegal activities. At that time, Global Witness concluded, "The problem lies with the capacity, the mechanisms to react to illegal activities, and the ability to deal with the armed forces, who continue to play the leading role in the destruction of Cambodia's protected areas."[10] Over the next decade, the RGC made many efforts to control illicit activities within the protected area system, though resources for management remained limited.

Social conflicts with communities that have been displaced by the creation of protected areas are apparent in Virechey National Park (VNP) and other protected areas. Initiatives to create community protected areas (CPA) as a way to engage resident villages in collaborative management systems are under discussion, though progress has been slow in formalizing policies and programs, and in securing financing, and implementing field activities. What has occurred at an accelerating pace is the transfer of use rights in Cambodia's protected areas to private companies. In 2012, for example, mining exploration rights have been issued for most of Virachey National Park, while leases to clear much of the eastern boundary with Vietnam for rubber plantations, totaling 16% of the park's area, have also been granted.[11] While Cambodian and international environmental groups protest the misappropriation of high-value protected areas for economic enterprises,[12] the Ministry of Environment has allocated 346, 000 hectares in 13 protected areas, representing 10.5% of the total park system, to rubber concessions, of which 251,000 hectares were granted in 2011 alone.[13]

There is little question that large-scale economic concessions granted in Cambodia's national parks and wildlife areas represent an even more serious threat than illegal logging. Rubber concession development requires both the complete clearing of natural forests as well as the development of roads and infrastructure that are associated with further development. This dramatic transformation of the landscape and ecosystem also fundamentally changes local economies and cultures. As an indigenous Kuy villager commented on a rubber plantation in Boeng Per National Park, "Since we have lost the resin trees and forest, we feel we have lost everything in our life."[14]

A combination of recent media reports documenting the rapid leasing of Cambodia's protected areas[15] and growing social conflict with resident communities has placed increasing public pressure on the government to reconsider current policies emphasizing economic development in conservation zones. The MOE, however, has few financial and human resources and limited political leverage to ensure that protected areas will remain protected from those who view them as targets for economic development.

FOREST CONCESSION MANAGEMENT FROM 1990–1999

Industrial timber exploitation was viewed as a key driver of economic development in Cambodia through the 1990s, despite a growing number of questions regarding its long-term viability. A report by the World Bank in 1996 explained that with over $1 billion invested and 40,000 employees, timber revenues contributed only 12 percent of GNP.[16] A paper produced by the Forest Policy Reform Project in 1998 noted that RGC revenues could be further increased, as "the full potential of the forestry sector to the economy is not yet realized."[17]

The Cambodia Forest Policy Assessment conducted by the World Bank projected an income of over $100 million per year by instituting higher royalty rates on timber concessions and enforcing sustainable use practices. High expectations for forest-based economic growth and sustainable management were common in many Southeast Asian countries during the 1970s, '80s, and '90s. Multinational and bilateral agencies provided financial and technical support to develop a sustainable logging industry, yet many countries were disappointed as unsustainable exploitation led to environmental degradation, social conflict, and revenue generation that fell far below expectations.

In Cambodia, a major driver of industrial logging was regional demand for timber from Thailand and Vietnam. In 1989, Thailand imposed a national logging ban in response to public protest over downstream flooding and other environmental impacts, and canceled over 300 timber concessions. The mixture of rapid economic growth in Thailand and other parts of Asia, combined with shrinking supplies of whole log exports, put immense pressure on Cambodia's forests for logs. In January 1994, Thailand agreed to stop log imports from Cambodia and close checkpoints, though illegal trade continued on a substantial scale. In addition, there was no capacity to halt illegal exports to Laos and Vietnam. By the mid-1990s, illegal logging was estimated to be approximately 1.5 million cubic meters per year.

Due to the expansion of commercial concessions and the rise in illegal logging, it was projected that deforestation would accelerate to 1 percent per year by 1993. According to the report, "The coastal and flooded ecosystems will practically disappear. In this scenario, total deforestation would be 1,900,000 hectares from 1993–2010."[18] In July 1998, in response to the deteriorating situation with existing timber concession management, the RGC issued a Declaration on Forest Management in Cambodia and abolished nine concessions, and over the next year, with support from the military, destroyed 784 illegal sawmills.[19]

SUSTAINABLE FOREST MANAGEMENT PROGRAM

The donor response to the declining state of forest resources, widespread illegal logging, and poor royalty recovery was to launch a series of sustainable forest management (SFM) initiatives that were intended to ensure a sustainable yield indefinitely while providing social and environmental services. The main projects included the World Bank's Forest Policy Reform Project (1997–1998), the Asian Development Bank's (ADB) SFM project (1999–2000), and the World Bank's Forest Concession Management and Control Pilot Project or FCMCPP (2001–2003). The projects were based on the assumption that SFM was the most effective management system for enhancing development and economic growth from Cambodia's forests. This belief came under increasing scrutiny. According to Bruce McKenney, who assessed the situation in 2002, "Although the reluctance of concessions to adopt SFM reforms is often attributed to a lack of knowledge about SFM practices and a need for government and concessionaire capacity building, reluctance toward reform reflects the incompatibility between SFM and concession profitability."[20]

A number of reports published under the Forest Policy Reform Project highlighted sector shortcomings, including the estimate that about 94 percent of timber production was illegal and figures showing that government timber revenues totaled just $6 million (1997) and $10 million (1998), despite having 6.5 million hectares under concession lease.[21] In spite of these problems, experts with the Forest Policy Reform Project concluded, "the forest concession system is the most appropriate mechanism for commercial development of forest resources in Cambodia, but needs refinement to suit Cambodian conditions."[22]

The ADB-financed Sustainable Forest Management Project found that the "extremely poor performance" of concessions indicated a "total system failure."[23] A report by Development Alternatives International (DAI) in 1998 estimated that Cambodia's forests would be economically depleted within five years if logging continued at the present levels.[24] While the World Bank continued to provide support for SFM, its own assessments seemed to question whether this management system would be profitable for concessionaires, noting that concessionaires would need to restrict themselves to harvesting no more than 10m³/ha versus the 50m³/ha or more that many concessions typically extracted. They concluded, "If limited to sustainable harvest levels, even with subsidized royalties, the current concessions are unlikely to be able to operate profitably as planned because of low conversion rates, poor market strategies, and high capital costs."[25]

As part of the donor-led effort to establish SFM policies and systems, the DF&W required concessionaires to prepare and submit sustainability plans and obtain approvals prior to continuing logging. Few concessions prepared SFM plans, and even fewer received approvals of their working plans. As a consequence, sanctioned field operations slowed towards the end of the 1990s. A 2002 assessment of financial *disincentives* to SFM included the following:[26]

- Rapid and intensive harvesting reduces concessionaires' risk exposure

- A far higher annual rate of return can be earned by harvesting intensively and investing profits elsewhere than by harvesting sustainably

- Concessionaires' existing investments in log processing capacity may be underutilized if SFM reduces logging harvests

IMPACT ON DEFORESTATION AND FOREST DEGRADATION

By 1997, it was estimated that only 6 percent of Cambodia's forests (625,177 hectares) had high commercial value—dense evergreen ecosystems. By contrast, 30 percent of the country's forest (3,183,000 ha) was classified as disturbed evergreen that had experienced moderate-to-severe extraction levels. Between 1973 and 1993, Cambodia's forest cover declined by 15 percent—a loss of approximately 2.1 million hectares of forest, representing an annual loss of 0.6 percent per year, with a 1.2 million hectare increase in degraded shrubland.[27] The deterioration and ultimate conversion of forestlands into shrublands often took place gradually. Initially, forests that were "high graded" had high-value species removed, either within approved concessions or through illegal felling. Timber concessions and illegal loggers typically overexploited forests, further depleting the forest structure and incurring substantial additional environmental damage due to extraction and road building operations. Subsequent fuelwood collection, charcoal manufacturing, and timber felling for building materials and other subsistence needs further degraded the ecosystem.

By 2000, the widespread loss of Cambodia's forests and the rapid process of forest degradation attracted international attention and was widely publicized by NGOs and the media. But, as land markets expanded rapidly over the next ten years (2000–2010), forests were cleared, burned, and claimed as private land that could be sold. Some logged-over forests were leased as economic land concessions and cleared for rubber, sugar cane, eucalyptus, and other plantation crops. Subsequently, land prices increased, such as in Oddar Meanchey Province and parts of the northeast.

FOREST REVENUE

Forest income was a major justification for leasing much of the country's forests to commercial timber interests. As a World Bank report (1996) notes, "Forests are one of the few publicly owned resources in Cambodia that have strong potential for making a significant contribution to badly needed growth in government revenues."[28] However, forest revenue data from the period demonstrate that the financial potential of forest income was never achieved. Despite estimates that forest concession management could yield $100 million per year in government revenues, annual forest revenue from 1996 through 2001 varied from $6 million to $12 million. With estimated extraction levels of 1.5 million cubic meters per year, the government was only able to recover 25 percent of revenue due from the harvest. While improved regulation and collection would have increased potential revenue, the adoption of SFM would have reduced production, raising further questions regarding the validity of high-revenue estimates.

A substantial number of policy studies were conducted to review price setting and income generation from the forest sector. Cambodia possessed richly stocked forests with some high-value species commanding market prices of over $500 per cubic meter in the 1990s. Royalty rates for concessions that averaged $14 per cubic meter represented less than one-fifth of the border price equivalent and were not indexed to international markets. The strategy behind keeping royalty rates low was that additional economic development benefits would be generated through employment and value-added processing.

Additionally, planners anticipated that logging concessions would invest in social and physical infrastructure projects, including schools, roads, and health clinics in areas under their control. These benefits failed to materialize, as timber companies sought to maximize profits and lacked the capacity and incentives to engage in rural development. A forest policy assessment report in 1996 recommended that royalty rates be raised to $74 per cubic meter.[29] Significantly, an analysis of concession profitability conducted by KPMG (Klynveld Peat Marwick Goerdeler) in 2001 found that five representative timber companies controlling 40 percent of the concession area in Cambodia had not generated (reported) any significant profits between 1997 and 1999, which also meant that no additional revenue was available to pay royalty fees.[30]

SOCIAL CONFLICT

The Cambodian forest concession system also had substantial negative impact on many rural communities. In most areas where concessions operated, communities complained that important trees and shrubs that produced non-timber forest products (NTFPs) were being destroyed in great numbers. High-value trees, including resin-producing species and trees that sheltered honey-producing bee hives, were being felled by loggers. Rural Khmer and ethnic minority households traditionally held clear usufruct rights to such valuable trees, though concessionaires frequently failed to recognize those rights.

In addition, valuable rattan and other canes were destroyed through logging and timber extraction. In forest-dependent populations, the resulting loss of income from NTFPs used for subsistence purposes and for sale significantly impacted household livelihood. For example, one study of resin tapping showed that the annual income from resin sales for four villages in Mondulkiri province was $61,000, with an average income per tree of $3.60 per year. This represents an annual household income from resin of $299–$377, a substantial proportion of total cash revenues for these families.[31] The felling of resin trees by timber concessions was consequently a major source of forest conflict, with frequent protests at the local and national level.

With 85 percent of the country rural and forest-dependent, and with nearly 60 percent of Cambodia's forests managed by logging concessions, considerable social conflict emerged which created tension at the local level and catalyzed a national debate on forest policy. Rapid deforestation and forest degradation also impacted the natural environment in numerous ways by exacerbating erosion and sedimentation of waterways including the Tonle Sap. Flood forests were estimated to have declined by 51 percent from 1970 to 1996, with a 19 percent loss in mangrove forests over the same period. Loss of biodiversity also accompanied the destruction of forest habitat. The rapid destruction of flood forests around the Tonle Sap and the coastal mangroves had significant impact on fisheries, as these ecosystems are spawning grounds for many fish species.

After reviewing the management plans of four Cambodian timber concessions, a German consulting firm concluded, "Social issues and impact of forest management in all four plans are inadequately dealt with. Village consultations and participation of communities in the selection and delineation of 'their' community forests are lacking and not following the guidelines."[32] In three of the concessions, social conflicts were so severe that the consultants questioned whether the concessionaires could find a solution that would allow the concessions to continue operating. Government, donors, and civil society groups agreed that action was needed to stop illegal logging, contain environmental destruction, and address social conflict. The consultant team noted that:

> Having seen… the extremely destructive effects of previous unscrupulous forest concession management, as well as the present extent of illegal logging, land grabbing, forest conversion, practically irreversible depletion of vegetation and soil resources, also seeing the underlying causes of social problems

and poor governance, we suggest…to allow and oblige communities to participate in management and protection of forests.[33]

In late 1999, Global Witness was contracted as an independent monitor
of the Forest Crimes Monitoring and Reporting Project (FCMRP) funded
by the World Bank. The goal of the initiative was to build government
capacity to detect, investigate, and control forest crimes. This initiative
was consistent with Prime Minister Samdech Hun Sen's directive to eliminate illegal logging, reflecting government concern over the acceleration
of illicit felling in the 1990s.

The World Bank's inspection panel noted that international and
national vested interests controlled logging concessions, and that the
capacity of the government to manage them and the extensive territory
under their control was limited. By 2000, one analyst described Cambodia's commercial timber industry as "a total system failure, resulting from
greed, corruption, incompetence, and illegal acts that were so widespread
and pervasive as to defy the assignment of primary blame."[34]

FOREST CONCESSION MORATORIUM

By the arrival of the new millennium, Cambodia's timber concessions
were in trouble and under growing pressure from government and third-
party monitors and regulators. There was growing recognition that the
forest concession development model had failed to generate revenues for
the government or to channel funds and stimulate development in rural
areas. There was also growing pressure on the Ministry of Agriculture,
Forests and Fisheries (MAFF) to cancel commercial timber concessions
that were not financially viable or that had failed to follow the terms of the
Sub-Decree on Forest Concession Management.

The downsizing of Cambodia's forest concession system was clearly
under way on January 1, 1999, when the RGC declared a moratorium on
logging. In that year, eight concessions were cancelled, totaling nearly 2.5
million hectares. In April 2000, a panel of experts involved in the ADB
concession review strongly recommended the adoption of the sustainable
forest management plan (SFMP) with performance-based indicators. The
report further noted that management plans should be submitted within

one year and that concessionaires who failed to meet even one indicator should have their concession cancelled.[35] By October 2000, the director general of DF&W and the Cambodia Timber Industry Association (CTIA) informed the World Bank that they accepted the SFMP, and that the concessionaires would need to have complied with ADB guidelines by November 2001 or face termination of their contracts.

According to Global Witness, in June 2002, the RGC announced the first cancellation of a major timber concession (GAT International). In fact, timber concessions had been cancelled and reallocated since 1994. Between 1994 and 2003, however, it is apparent that the area under concession management declined substantially, from around 6.5 million hectares at the peak to 2.7 million hectares in late 2005, reflecting a shift away from commercial timber management (see table 1).

Table 1. Concession Cancellation or Reallocation 1994–2005

Year	Concessions Cancelled or Reallocated	Area (ha)
1994	2	232,374
1996	6	936,841
1997	3	396,960
1998	3	2,137,960
1999	8	2,470,446
2000	4	363,724
2001	No Cancellations	0
2002	3	448,700
2003	3	1,171,782
2004	No Cancellations	0
2005	No Cancellations	0
Total Concessions Cancelled or Reallocated	32	8,158,787
Concessions remaining as of November 2005	13	2,713,828

Source: Inspection Panel, Cambodia: Forest Concession Management and Control Pilot Project (Washington, DC: World Bank), March 30, 2006, 110.

The closure of the GAT International concession was just part of a process of discrediting SFM and its supporters, including the World Bank. On January 28, 2011, a letter requesting a review of the World Bank's Cambodia FCMCPP was submitted to the Bank's Inspection Panel. NGO Forum and Oxfam Great Britain accused the World Bank of not complying with its own policies and procedures by promoting the interests of logging concession companies over those of the people. The complaint noted:

> By assisting the concessionaires in producing forest management plans and environmental and social assessments, the Bank used loan money to benefit logging companies with track records of, among other things, human rights abuses and illegal logging of resin trees, which are one of the main sources of livelihood of forest-dependent communities.[36]

The failure of timber concessions to produce sustainable forest management plans resulted in both cancellations and an ongoing moratorium on timber operations. The collapse of the industrial timber management system created a growing forest management vacuum in the field, placing growing pressure on government to find new management systems for the nation's forests. Over the past decade, several strategies have received growing interest from the MAFF, including community forestry, decentralization, and social and economic land concessions.

FOREST SECTOR TRANSITION 2000–2010

Downsizing of the forest concession system in Cambodia created both a need and an opportunity for alternative approaches to forest management. In 2004, a joint donor and government Independent Forest Sector Review endorsed an end to the timber concession system, suggesting commune-level forest partnerships as an alternative.[37] In November 2005, the World Bank reported to the RGC that their findings indicated that the behavior of concessionaires was not likely to improve and that the government should exercise its right to cancel concessions and adopt other strategies, including community forestry and commune-based forest protection.

By 2007, all forest concessions were inactive and a growing number had been cancelled.[38] This situation presented policymakers with a number of options for moving forward. While this book will focus on transitions to community forest management systems, some other approaches will also be discussed.

ECONOMIC LAND AND MINING CONCESSIONS

Over the past decade, it is apparent that the suspension or cancellation of logging concessions has often resulted in a shift of leases to agricultural and mining concessions. In 2006, it was reported that more than 943,069 hectares of land had been allocated for agricultural concession, though it is not known how much of this land was originally under forest concession.[39] By 2012, a Cambodian NGO estimated that nearly 4 million hectares of land had been granted in the form of mining exploration rights and economic land concessions (ELC) representing 22 per cent of Cambodia's surface area.[40] Of this area, 2 million hectares was leased to 227 agro-industrial companies for rubber, sugar cane, and cassava plantations, with another 1.9 million to mining companies for exploration rights. In addition, it is reported that an estimated 30% of the economic concession leases are held by a small group of Cambodian senators and businessmen.[41] The Ministry of Agriculture put the figure at 1.2 million hectares allocated to 118 agro-industrial firms, including 28 Chinese companies and 27 Vietnamese firms. The issuance of ELCs has been criticized for the lack of transparency and reporting reflected in the large discrepancies in data generated by government agencies and civil society. This has contributed to Cambodia being ranked 164th out of 182 countries in terms of perceptions of corruption.[42]

It is also apparent that many of the former forest concessions that were depleted of timber are being converted to large-scale plantations for rubber, sugarcane for biofuels, and pulpwood (eucalyptus and acacia). Forest conversion into agricultural uses has also fueled the commoditization of land, with "land grabs" becoming increasingly common in heavily forested provinces such as Ratanakiri and Oddar Meanchey. Economic land concessions (ELCs) have been criticized from a human rights perspective for their failure to respect the resource rights and livelihood needs of resident people. The UN report notes that:

> At the root of these concerns is poor enforcement of and compliance with the requirements of the Land Law and Sub-Decree on Economic Land Concessions, which govern the granting and management of the economic land concession."[43]

The UN report concludes that:

> Economic land concessions have not led to increased agricultural productivity or economic growth in Cambodia, and large areas of conceded land have been left idle or underutilized."[44]

The UN report recommends that the RGC adhere to existing laws that require land and natural resources to be managed in ways that benefit all Cambodians, with special protection required for forest-dependent and indigenous communities.

In 2008, the Sub-Decree on Economic Land Concessions was amended to guide the allocation of agro-industrial concessions, including rubber, oil palm, and fast-growing timber plantations, as well as animal husbandry and other large-scale agricultural production activities on state lands. The land must be delineated, demarcated, and registered as state private land. Environmental and social impact assessments must be completed with public consultations with local authorities and resident communities. While the ELC issuance process has been modified, requests for ELCs often compete with applications by communities for CF agreements, as well as by those of indigenous communities seeking recognition of ancestral domain claims. Coordinating and balancing state public land allocation in ways that respond to national environmental and social goals, while also providing opportunities for legitimate economic development investments, remains a challenge.

The acceleration in issuing economic land concessions in 2010–2011 has limited the capacity of the government and civil society to negotiate and assess legitimate community claims to forest lands. As a consequence, many areas that have traditionally been held as community forest use areas, as well as new areas with potential for community forest management, are being allocated to private sector, and often foreign-held firms.

As mentioned earlier, ELC and mining exploration permit issuance has generated growing social conflict, donor concern, and civil society protests that have recently resulted in Cambodian Prime Minister Hun Sen's decision to place a temporary suspension on any new ELC and on any evictions of local people from concession areas. The recent crisis over ELCs makes it increasingly clear that Cambodia's contested forest domain represents a critical area of policy debate with decisions regarding what forest lands will be allocated to communities and local governments and what will be leased to private firms, gaining greater political visibility and risk for the national leadership.

SOCIAL LAND CONCESSIONS

The RGC has sought to address problems of landlessness in the Lower Mekong Region through a program of Social Land Concessions (SLC) that has been ongoing over the past decade. According to recent reports, 20% of rural households in Cambodia are landless and 75% own less than one hectare of land. In 2011, 12.1 million people inhabited the more densely populated Lower Mekong Region with an average population density of more than 200 per km^2; by contrast, 1.3 million people resided in the peripheral districts with a population density of less than 15 inhabitants per km^2. The Social Concession program sought to assist landless and land-poor families from the core area by identifying agricultural lands in the periphery where they could resettle.[45]

The 2003 Sub-Decree on Social Land Concessions (SLC) sought to reduce poverty and improve land management, while strengthening the authority of the commune councils. The sub-decree outlines criteria, mechanisms, and procedures for granting residential use and family farming rights. Allocated land can become privately owned and titled provided that the recipient follows defined criteria within a five-year period. Commune councils, however, must follow a complicated process to create plans for an SLC, including identifying land and participants, conducting social and environmental impact assessments, assessing infrastructural requirements, and assuring ongoing management and monitoring.

By 2011, only 6,277 ha. of SLC had been granted in contrast to 1.2 to 2 million ha. of ELC.[46] Some 1614 households had participated in the program, in contrast to over 100,000 landless families that have been targeted for the 1 to 3 hectare concessions. A challenge confronting the project is the identification of available land for SLCs. A 2004 survey in 40 communes in 8 provinces with low population levels sought to identify available land for SLC. After a broad-based stakeholder dialogue with commune chiefs, village leaders, elders, and community members, it was found that most communes had little land available for SLC and that those were of poor quality.[47]

By contrast, ELCs seem to have experienced little difficulty having lands allocated and are outpacing SLC issuance at a rate of 200–300 ha. to 1 ha. A recent analysis of the competition between ELCs and SLCs suggests that ELCs have national and international investor backing with a strong financial base, political support, but a non-transparent process to secure land, while SLCs have no financial resources, weak or no political support, and a difficult process to follow.[48] This experience seems to suggest that the vision of the Land Policy Declaration of 2009 is not being achieved when it states that "Land distribution shall ensure equity, social stability, food security…for sustainable development, prevent land concentration and promote productive and effective use of the land."[49] Again, this situation reflects the nature of Cambodia's contest forest domain.

DECENTRALIZATION

Over the past decade, policy makers and donors have been exploring approaches to increase the role of commune councils in natural resource management as part of a broader national decentralization process. This will require meaningful transfers of discretionary power and authority to local institutions that are accountable to the local populations. In Cambodia, as in some other countries, this process of devolution can be problematic, in part due to reluctance on the part of the existing authorities to transfer forest tenure rights to local groups. Nonetheless, decentralized natural resource management could improve the sustainability of resource use and increase social equity, and is also a priority in Agenda 21 of the Rio Declaration and the Millennium Goal Declaration. In Cambodia, the decentralization strategy involving local government in forest management is known as *partnership forestry.* This strategy was recommended by the donor-sponsored Internal Forest Sector Review (IFSR) in 2004 and supported through a 60-million-dollar DANIDA/DFID livelihood program implemented in Cambodia from 2006 to 2010.[50]

In Cambodia, there are 1,621 communes (*sangkats*), each with three to fifteen villages, with a total population of 15,000 to 20,000. Under the proposed partnership forestry initiative, commune councils would be required to prepare forest management plans and to implement them after approval from the Forestry Administration. A recent analysis identified several issues that have constrained implementation of partnership forestry through commune councils. These include weak commune councils with no discretionary powers in forest management, unclear relationship between commune councils and community forestry groups, and limited willingness of the Forest Administration to devolve management authority to the commune council.

Commune councils are in a difficult position to implement management activities. They are frequently too far from the center to attract financial resources for development activities, and too far from the forest to be actively engaged in direct management. Potentially, they could help coordinate the efforts of village-based CF groups and provide a formal interface with national government. But, the respective roles and responsibilities of national forestry agencies (FA, MOE, etc.), local governments, and communities require greater clarity and support for decentralization to move forward. The decentralization research team concluded that:

> In Cambodia, devolution . . . is unlikely to work unless the communities get access to substantial benefits from the forests, and unless they are enabled to deal with strong external and internal actors and conflicting interests.[51]

COMMUNITY FOREST MANAGEMENT

NGO-supported community forestry pilot projects emerged in the early 1990s, with a national CF working groups, involving government and civil society, meetings by 1997. Through a process of community and civil society consultations, a national community forestry movement emerged. In 2000, the Department of Forests and Wildlife (DF&W), currently reorganized as the Forestry Administration (FA), was under the scrutiny of foreign development agencies and the mass media. There was increasing pressure on the government to shift from awarding logging contracts to emphasizing forest restoration, conservation, and community benefits instead. In response, the DF&W developed a national community forestry strategic plan to guide Cambodia's efforts to formally engage and recognize the rights of forest-dependent villages as partners in public resource management. The original vision sought to distribute forest benefits in a sustainable and equitable way.[52]

The plan was developed by a diverse team of young professional Cambodian foresters, foreign advisers, donor agency staff, and NGO representatives. It drew heavily on Cambodian field experiences with community forestry pilot projects that had been implemented by two international NGOs—Concern Worldwide, affiliated with the Mennonite Central Committee, and the UN Food and Agricultural Organization (FAO). The plan initiated the implementation of fifty-seven community forests (CFs) in Cambodia, putting approximately 83,000 hectares under community forest management.[53] These early experiences with community forestry generated a wealth of experience and an initial "proof of concept" of this strategy.[54] Over the next decade, a national strategy, policy, and legal framework emerged to establish CF as a fundamental approach to state public lands management.

While a government-sponsored CF initiative was under way, forest areas under traditional management implemented by indigenous minority communities and Khmer villages were engaged in their own informal forest management activities. These initiatives were not connected to government projects or directives. The challenge facing the RGC was how to formalize community efforts to protect their forests and sustainably manage them. However, creating an enabling framework that allows informal and traditional resource management systems to mesh well with formal state-sponsored systems was challenging. At the same time, without formal recognition, CF groups would lack clear tenure security and usufruct rights, potentially creating conflict with authorities and limiting access to technical and financial resources.

The RGC held stakeholder consultations and integrated field learning into the design of its CF policies, legal framework, and implementation strategy while recognizing a need for flexibility and adaptation in integrating informal community forestry systems into the national forest program. In 2003, the RGC approved national legislation with the Community Forestry Sub-Decree, and in 2006 it approved Implementation Guidelines (*Prakas*) for Community Forestry. These two actions provided both field operational guidelines for CF and a national umbrella program that includes community forestry, community-based production forestry, partnership forestry, and community conservation forestry.

In 2006, the FA developed the National Community Forestry Program (NCFP) as one of six components under its National Forestry Program (NFP). The NCFP mandated the identification and formalization of CF groups, with support for community institutional and livelihood development.[55] By 2011, 389,021 hectares of forest area was being managed by 441 CF groups that have either been legally recognized or are in the process of securing recognition by the FA. By 2020, the FA is seeking to further extend community forestry to cover two million hectares with one thousand community forestry management committees (CFMCs) established.

GOALS AND POTENTIAL

The goal of the NCFP is to broaden programs and policies to cover all forms of decentralized forest management modalities.[56] Community forestry is viewed as a fundamental strategy to enhance rural livelihoods and reduce poverty through the provision of direct and secure access to natural resources. This community forestry vision assumes that by creating tenure security and enhancing forest livelihoods, forest-dependent villages and households will have strong incentives to invest in sustainable forest management.

Over the next decade, the FA has set a target of transferring two million hectares of state public forests to communities, a transfer representing approximately 20 percent of the nation's forests. Whether the FA and its partners will be able to achieve this ambitious goal remains to be seen; however, progress made in recent years suggests a firm commitment to further devolution. An assessment made in 2002 estimated that up to "8.4 million hectares of Cambodia's forestland could be suitable for CF. In addition, 3.9 million hectares of agricultural land could be managed under Farm Forestry."[57] The 2002 CF assessment came in with high figures, as it was based on forest area within 10 km of rural communities and did not consider jurisdiction, user rights, or concession boundaries that would be important in determining potential areas for CF. Nonetheless, the study is informative in indicating the extent of community forest dependence in Cambodia (see table 2).

Table 2. Forest Buffer Areas within 10 Kilometers of Communities in Cambodia by Forest Condition[58]

Category	Hectares within 10 km Buffer Zone	Percentage of Cambodia's Forest Cover	Population
Good Forests	5.5 million ha	51%	299,024
Degraded Forests	2.3 million ha	21%	622,195
Mangrove and Flood Forests	0.6 million ha	6%	144,605
Total	8.4 million ha	78%	1,065,824

The NFP mandates, "All forest-dependent villagers living in and near the forest are given the opportunity to be involved in legally recognized community forestry management of some local forest…"[59] The amount of forestland that eventually comes under community management is determined by a number of factors, which include the following: the availability of financing from government and donor agencies to build community capacity for implementing the NCFP; existing concessions and jurisdictions; competing future demands for forestland by other government agencies, the private sector, and the military; and the extent of political and policy support for the program. Over the past decade, community forestry has made great progress in terms of policy and legal endorsements from government and expanding financing from development agencies. If this trend continues, the area under CF could expand rapidly.

Important determinants of CF's long-term success in Cambodia will be its coverage and impact on the ground. Key components will include how effective rural CF groups are in protecting local forests, the degree to which they are able to generate increased income flows from forest resources, and the extent to which existing CF groups encourage neighboring communities to organize and initiate forest protection. At the same time, while *demand* for CF in rural communities may increase, the capacity of the FA and supporting NGOs to *supply* technical and financial support will also determine the rate at which the NCFP can be implemented.

IMPACT AND COVERAGE

By September 2011, the NCFP was monitoring and supporting 441 CF groups nationwide, with a total coverage of 389,021 hectares, or approximately 4 percent of Cambodia's forest area. The development of a CF group, according to the *Prakas*, includes the steps described in subprogram one. The key step in securing government recognition, however, is the seventh step, "preparation and approval of the CF agreement." Typically, this step involves processing CF requests at the FA Inspectorate, then forwarding that application to FA headquarters, followed by approval by the Ministry of Agriculture, Forestry, and Fisheries (MAFF). The process takes several months to several years depending on the quality of the application, the motivation of local FA staff to review the proposal, and the rate at which action is taken by national agencies.

Once the CF group application is approved by MAFF, a formal agreement can be signed. By September 2011, 61 percent of the 441 CF groups under development had been approved by MAFF, and 34 percent had signed agreements with the FA. The majority of the CF group approvals have been issued in the North Tonle Sap Inspectorate, where active collaboration between local FA and FAO, Community Forestry International, Pact, and a host of other NGOs have led to 83 percent of CF groups being formally recognized.

Some cantonments and provinces have moved more quickly than others to secure approval from MAFF, while some have yet to achieve approval. For example, in Kompong Thom Province, FA cantonment staff worked closely with the CBSFLMP team (Capacity Building for Sustainable Forest and Land Management Project) to formalize fifty-two

CF agreements by 2011, representing nearly 45,499 hectares of CF areas and involving sixty-three villages.[60] This experience demonstrates how motivated FA field staff with adequate technical and financial support can accelerate NCFP implementation. By contrast, in Ratanakiri Province, only 4 percent of the 20,699 hectares of community forests currently under development have been approved by the FA. In part, this may be due to the extensive allocation of concessions in this region that needs to be addressed prior to approval, the lack of initiative on the part of local FA staff, and limited capacity to work with ethnic minority communities that reside in the area.

The amount of forestland under community forestry in Cambodia was 4 percent in mid-2010; however, some forest-poor provinces like Takeo had as much as 68 percent under CFMCs. In Kompong Thom and Oddar Meanchey Provinces, where CF support programs have been ongoing for some time, 13 percent and 15 percent of state public forests are under community control). The size of CF forests also varies widely, from Oddar Meanchey with an average of 4,920 hectares to Pursat with 117 hectares per CF group. Size typically reflects a combination of the extent of forest cover and population density. The South Tonle Sap Inspectorate, with high population and low forest area, tends to have much smaller CF holdings. This requires CF groups to adopt different strategies and management activities. Typically, provinces with high population densities and less forest cover also tend to possess more degraded forests.

FUTURE CHALLENGES AND OPPORTUNITIES

There has been growing recognition among government planners, development agencies, NGOs, and researchers that forests provide valuable environmental and socioeconomic services. In Cambodia, rural poverty has declined from 43 percent of households in 1994 to 34 percent in 2004.[61] In remote rural areas where village households depend heavily on forest resources, the number of families living below the poverty line may be considerably higher, especially if forests have been degraded due to logging or land clearing. In such areas, one study indicated that NTFPs contribute up to 60 percent of household income.[62] Another report found that forests in Kompong Thom and Kompong Chhnang provided up to 100 percent of the livelihood requirements of poor, landless households, and 75 percent of the income of poor farmers.[63] While alternative management systems such as estate crops under ELCs may generate higher export revenues and foreign investment that may foster national economic growth, the costs to forest-dependent communities who lose these resources are substantial, as are the hard-to-quantify environmental costs as ecological services are lost as forests are cleared.

The RGC's five-year National Strategic Development Plan 2006–2010 (NSDP) acknowledges the need to improve the livelihoods of the rural poor as a top priority. A key recommendation of the World Bank's Cambodia Poverty Assessment (2006) was to provide "equitable access to common property resources as a critical source of income and security for the rural poor."[64] The report notes that 25 percent of the poor, on average, depend on fishery and forestry products for over half their income. At the same time, access to forests and common property has been shrinking as forests are allocated to concessions, captured by land speculators, or degraded through unsustainable use. The report noted that:

> Numerous studies of rural livelihoods paint a consistent picture of declining access to these resources; especially for the poor... part of it is due to a real decline in resource availability, as population increase and commercial interests have driven unsustainable overexploitation.[65]

A challenge the government faces is how to transition the forest sector into sustainable systems of management that provide important environmental and social benefit, while generating employment and income for rural, forest-dependent communities. The RGC has developed an elaborate law and policy framework to support resource governance over the past twenty years, with many components supportive of community management including fisheries, forestry, and land, which the Community-based Natural Resources Management Learning Institute and Melissa Marschke have summarized below (see table 3).

Table 3. Cambodia's Policy Framework for Resource Governance[66]

Resource Area	Resource Governance Policy
Protected Areas Policy	■ Royal Decree on the Creation and Designation of Protected Areas: 1993 ■ Environmental Protection and Natural Resources Law: 1996 ■ National Environmental Action Plan: 1998–2002 ■ Protected Areas Law: 2008 ■ Guidelines on Community Protected Areas Management: In progress
Land Policy	■ Land Law: September 2001 ■ Sub-Decree on Procedure for Commune Land Use Planning: 2008 ■ Draft Declaration on Land Policy: In progress ■ Sub-Decree Procedures of Registration of Land of Indigenous Communities: 2009
Fisheries Policy	■ Fisheries Policy Reforms: 2000–2001 ■ National Fisheries Policy Statement: 2005 ■ Fisheries Law: 2006 ■ Sub-Decree on Community Fisheries Management: 2005, 2007 ■ Guidelines for Community Fisheries: 2007 ■ The Strategic Planning Framework for Fisheries: 2010–2019
Forestry Policy	■ National Forest Policy: 2002 ■ Forest Law: 2002 ■ Community Forestry Sub-Decree: 2003 ■ Guidelines for Community Forestry: 2006 ■ National Forestry Program: 2010 ■ National Community Forestry Program: 2011

Local Governance Policies	■ Law on Commune Administration: 2001
	■ Organic Law on Sub-National Administration: 2008
Other supportive Policies	■ National Poverty Reduction Strategy: 2003–2005
	■ National Strategic Development Plan: 2006–2010
	■ National Strategic Development Plan: 2009–2013

The emergence of a National Community Forestry Program, supported by core legislation over the past decade, and the accelerating rate of implementation discussed later in this book provide evidence that national planners are responding to this need and this opportunity. At the same time, the NCFP necessarily competes with other RGC development goals and programs. The extent to which the NCFP succeeds in achieving its goal to devolve forest management rights to communities covering two million hectares will depend both on capacities to implement the program, as well as the political will to do so. Over the next decade, the priority the RGC will give to the NCFP and other devolution policies and programs versus commercial development will likely influence the future of Cambodia's landscape and society.

NOTES

1. A. Mittelman, "Secondary Forests in the Lower Mekong Subregion: An Overview of their Extent, Roles and Importance" *Journal of Tropical Forest Science* 13, no. 4 (2001): 671.

2. Ibid., 674.

3. Ibid., 676.

4. The World Bank, "Forest Policy Assessment: Cambodia" (Washington, DC: The World Bank, 1996).

5. Bruce McKenney, "Questionable Sustainable Concession Forestry in Cambodia" *Cambodia Development Review* 6, issue 1 (January–March 2002): 1.

6. See *Cambodia: Forest Policy Assessment* (World Bank/FAO/UNDP, April 29, 1996).

7. David A. Ashwell, *Cambodia: A National Biodiversity Prospectus* (Phnom Penh: IUCN, 1997).

8. Collin MacAndrews, "Strengthening Institutions for Implementation of Forest Policy in Cambodia" (Phnom Penh: ARD, Inc., 1998), 11.

9. Ibid., 4.

10. Paul Vrieze and Kuch Naren, "Carving Up Cambodia One Concession at a Time," *The Cambodia Daily* (March 10–11, 2012), 11. See *Deforestation without Limits* (London: Global Witness, 2002), 18.

11. Vrieze and Naren, "Carving Up Cambodia," 11.

12. Ibid.

13. Ibid., 7.

14. Ibid., 10.

15. Ibid., 5–11.

16. See *Cambodia from Recovery to Sustained Development* (World Bank: Washington, DC, 1996). All monetary amounts are listed in US dollars, unless specifically stated otherwise.

17. MacAndrews, "Strengthening Institutions for Implementation," 3.

18. The World Bank, "Forest Policy Assessment: Cambodia" (Washington, DC: The World Bank, 1996).

19. *National Community Forestry Strategic Plan* (Phnom Penh: Department of Forests and Wildlife, 2000).

20. McKenney, "Questionable Sustainable Concession Forestry," 2.

21. Ibid.

22. *Findings and Recommendations of the Log Monitoring and Log Control Project: Main Report* (Bethesda, MD: DAI, 1998), submitted to the Ministry of Agriculture, Forestry and fisheries, Royal Government of Cambodia).

23. McKenney, "Questionable Sustainable Concession Forestry," 2.

24. DAI "Findings and Recommendations of Log Monitoring and Log Control Project: Main Report" (Bethesda, MD: Development Alternatives, Inc., 1998).

25. The World Bank, "Forest Policy Assessment: Cambodia" (Washington, DC: The World Bank, 1996).

26. McKenney, "Questionable Sustainable Concession Forestry, 3–4.

27. See Jim Carle, "Reforestation and Natural Forest Rehabilitation Policy in Cambodia" (Forest Policy Reform Project, Technical Paper No. 3, 1998).

28. Ibid., 50.

29. Ibid., 23.

30. *The Equitability of the Forest Taxation System in Cambodia* (Phnom Penh: KMPG International Forestry and Environmental Advisory Services Group, 2001). Prepared for the Cambodia Timber Industry Association.

31. D. Tom Evans, Hout Piseth, Phet Phaktra, and Hang Mary, *A Study of Resin Tapping and Livelihoods in Southern Mondulkiri, Cambodia with Implications for Conservation and Forest Management* (Phnom Penh: World Conservation Society, 2003).

32. Hohannes Huljus and Britta Jell, *Cambodia: Review of Strategic Forest Management Plans Prepared by Concession Companies Operating in Cambodia, Part II* (Hamburg, Germany: GFA Consulting Group, 2005), 11.

33. Ibid., 17.

34. Thomas Fraser, GFA-AGRAR, ANZDEC, "Cambodia Forest Concession Review Report" (Sustainable Forest Management Project, Asian Development Bank [TA- 3152-Cam], Ministry of Agriculture, Forestry and Fisheries, General Directorate of Forestry, Royal Government of Cambodia, Phnom Penh, 2000), v.

35. *Deforestation without Limits* (London: Global Witness, 2004), 4.

36. *Cambodia: Forest Concession Management and Control Pilot Project* (Washington, D.C.: World Bank Inspection Panel, 2006), x.

37. *Independent Forest Sector Review* (Phnom Penh: DFID, 2004).

38. *Community-based Forest Protection: An Option for Cambodia* (Phnom Penh: NGO Forum, 2007).

39. Ibid.

40. Vrieze and Naren, "Carving Up Cambodia," 9.

41. Ibid.

42. http://cpi.transparency.org/cpi2011/results/#CountryResults.

43. *Economic Land Concessions in Cambodia: A Human Rights Perspective* (Phnom Penh: UN-HCHR, 2007), 2.

44. Ibid., 22.

45. AFD-Cambodia, *"Case Study on the Links between Migrations, Agricultural Expansion, and Deforestation in the Pailin Province"* (AFD, Jérémie Dulioust, Mai 2011).

46. GIZ, "Land Distribution in Cambodia—Experiences and New Approaches for State Land Management" (FIG Working Week 2011, Marrakech, Morocco, 18–22, May, 2011). http://www.fig.net/pub/fig2011/ppt/ts04b/ts04b_mueller_5376_ppt.pdf.

47. Khorn Dinravy and Andreas Groetschel, "Poverty and Social Impact Analysis of Cambodia's Proposed Social Land." Workshop Documentation. May 20, 2004, Phnom Penh, Cambodia. http://siteresources.worldbank.org/INTCAMBODIA/Resources/PSIA-Documentation-And.pdf.

48. GIZ, "Land Distribution in Cambodia," 5.

49. Ibid.

50. Iben Nathan, Tove E. Boon, Sovatha Ann, and S. Vanny, "Constraints and Options in Local Forest Management in Cambodia, Is Decentralization a Solution?" (Unpublished manuscript, 2009), 7.

51. Ibid.

52. "Final Draft National Community Forestry Strategic Plan" (Phnom Penh: Department of Forestry and Wildlife, 2000).

53. Ibid., 5.

54. Jurgen Fichtenau, Ly Chou Beang, Nup Sothea, and Dy Sophy, "Assessment of Ongoing Community Forestry Initiatives in Cambodia" (Phnom Penh: DFoW/GTZ, 2002).

55. "Cambodia's National Forest Programme" (Phnom Penh: Ministry of Agriculture, Forestry and Fisheries, 2010).

56. Ibid., 71.

57. Fichtenau et al., "Assessment of Ongoing Community Forestry," 5.

58. Ibid., 50.

59. "Summary of Project Achievements, Lessons Learned, and Recommendations on the Implementation of the Japan Social Development Fund Grant for Capacity Building for Sustainable Forest and Land Management" (RECOFTC, 2010), 21.

60. "Summary of Project Achievements," 13.

61. *Cambodia: Halving Poverty by 2015* (Phnom Penh: World Bank, 2006), i.

62. NGO Forum (2007) Community-based Forest Protection: An Option for Cambodia (Phnom Penh: NGO Forum).

63. *Access to Forest Resources and Landlessness: Case Studies of Degraded Forests and Livelihoods in Kampong Thom and Kampong Chhnang Provinces* (Oxfam GB Cambodia: Land Study Project Mini Case Study, 2002), 2.

64. *Cambodia-Poverty Assessment 2006* (Phnom Penh: World Bank, 2006), ii.

65. Ibid., ix–x.

66. Table was developed by Melissa Marschke and is based on data collected and analyzed by CBNRM-LI, 2009.

PART II

COMMUNITY FORESTRY
IN THE FIELD

COMMUNITIES AND FOREST MANAGEMENT

MARK POFFENBERGER
KATHRYN SMITH-HANSSEN

CAMBODIAN COMMUNITIES, BOTH KHMER AND INDIGENOUS
ethnic minorities, have managed the country's forests for centuries, sustainably utilizing these natural resources to meet their subsistence needs and
for small trade. In the 1970s and 1980s, many indigenous forest management systems were disrupted during the civil war and period of social
instability, when warring factions financed their violence through log
sales, mostly to Thailand. Despite civil war, population dislocations, and
widespread logging, community forestry activities persisted and expanded
as stability returned to Cambodia in the early 1990s.

Yet, with the expansion of large timber concessions communities
faced new threats to their local forests. Social conflicts increased as forests
were degraded and cleared and income-generating resin trees were felled.[1]
An FAO Forest Resources Assessment estimated annual deforestation
in Cambodia at 131,000 ha from 1981 to 1990.[2] Logging, agricultural
expansion, charcoal manufacturing, and fuelwood collection had an even
greater impact in modifying Cambodia's forest ecosystems as they tran-
sitioned from high density forests to medium to low density forests and
forest fragments. According to one study, by 1993, 9.8 million ha or 54%
of the total land area was under low to moderately dense forest cover, with

only 1 million ha or 5.8% under high density forest cover.[3] These changes in forest structure, biomass levels, hydrological functions, and biodiversity, largely resulting from human-driven disturbances, have likely negatively impacted forest-dependent communities in recent decades. Aside from the loss of valuable resin trees, the availability of other non-timber forest products may also decline as forest ecosystems are degraded, with additional problems with spring water availability and stream flows emerging. Unfortunately, there have been few longitudinal studies in Cambodia regarding how forest degradation processes affect forest communities.

Since the moratorium on timber concessions was enacted in 2000, pressure from logging concessions decreased, though illegal logging remains a problem. However, the issuance of logging concessions has increasingly been replaced by the awarding of Economic Land Concessions (ELCs) which not only cut valuable trees, but often clear-cut forest lands for commercial crops. Often covering tens of thousands of hectares, ELCs represent the most serious threat to community forests as it can only be controlled and regulated by national policy makers. Although there are other threats to community forest lands such as illegal logging, soldiers, and landless migrants, these threats have motivated a growing number of villages to establish new CF groups or strengthen the capacity of traditional management bodies.

As Chapters 6 and 7 illustrate, communities may view the tenurial status of their forest resources in a very different way than the RGC. Many Khmer and indigenous ethnic communities perceive the forest as a place of the Buddha and forest spirits, a resource held in trust for the local needs of future generations, and view their role to act as community stewards. One study asked communities in three provinces who they believed owned the forests in areas where they collected forest products. They found that 75% of the households felt the forests were owned by the local villagers or were common property open to all, while only one-quarter acknowledged that the state owned the forest.[4] Almost all respondents were aware that the state claims the forest under the law; however, they dispute "the basis of this claim, arguing that by living in or near to the forest, and having used the forests resources for decades, they should be the rightful custodian of local forests."[5]

By contrast, the RGC recognizes these resources as the public forest estate to be utilized for national development. Over the past two decades, the RGC has allocated a majority of the forests for commercial purposes, first to timber concessions and later to ELC. As one Cambodian legislator notes, "The big increases in ELC are made in compliance with the government's development policy in alleviating poverty for the country."[6] These two diverse perspectives on the goals of forest management and resource rights have generated frequent conflict between stakeholders on both sides.

The opportunity to bridge this gap and resolve conflict is being attempted through integrating traditional forest management practices and institutions into the evolving National Community Forestry Program (NCFP). This process is complex and lengthy as discussed in chapter 1, and while civil society and the RGC have made considerable progress in formalizing CF groups over the past five years (2007–2012), approximately ten to twenty times as much forest area has been allocated to ELCs when compared to the area allocated to CF groups. It is apparent that senior policy makers within the RGC who support the broad-based implementation of CF agreements targeting 2 million hectares by 2020, and who want to retain 60% of the nation's land area under forests, must compete with other senior government officials who contend that much of this land should be allocated to ELCs to achieve development goals and often their own personal gain.

In Cambodia, as in many other nations in Southeast Asia, policy makers have generally prioritized the allocation of state forests to industry and commercial interests in order to support high rates of economic growth, rather than to communities to achieve livelihood goals and more sustainable forest management. While economic contributions from timber and economic concessions in reducing poverty have often been disappointing, other impacts such as high rates of forest loss and degradation, hydrological disruption, socio-economic disruption, and social conflicts are readily apparent. The accelerating issuance of ELCs and mining exploration rights by the Ministry of Environment (MOE) and the Ministry of Agriculture, Forestry and Fisheries (MAFF) in recent years appears to have intensified social conflicts between communities, concessionaires, and government to such an extent that the Prime Minister placed a temporary moratorium on ELC issuance in May 2012. Whether this indicates a more balanced policy approach that will increase allocations of forest land to CF groups remains to be seen.

CONTESTED FORESTS

Most of Cambodia's forests could be considered "contested domain," in terms of the different perspectives of stakeholders, especially those of government and local communities, regarding their respective usufruct and ownership rights. Assessing the amount of Cambodia's forests under traditional or more recent forms of community management is a difficult task. In 2002, the Cambodian-German Forestry Project estimated the potential for CF initiatives by including all forests within a 10 km walking distance from each rural community. The study found that 8.4 million hectares of potential CF was available[7] under this scenario, representing approximately 80% of the nation's forest area.

A 2004 study, by the Wildlife Conservation Society (WCS) and the Cambodia Research Institute (CDRI), found that 2,000 villages with 1.4 million people, or approximately 12 percent of Cambodia's population, were located within 5 km of high-value evergreen or semi-evergreen forests.[8] If dry deciduous and lower-value forests are included, up to 70% of Cambodia's population may rely on forest for non-timber forest products (NTFPs) to sustain their livelihoods. A study by Oxfam GB found that 26% of households in the three study districts collected NTFPs to support their incomes.[9]

In Ratanakiri Province, forest management has been practiced for generations following customary laws and traditions of a number of cultural communities. The half-dozen indigenous populations that reside in the area have used the forests as a pool of land for long-rotation agriculture. Indigenous people have dozens of complex classifications for their local forests including sacred groves, bamboo forests, watershed and spring forests, as well as a large body of forest land that may be cleared periodically for agriculture, used for several years, and then left to regenerate. One such system is described in chapter 7. While land registration is just beginning to take place, and communal tenure has recently been recognized under the Land Law, there is still no official data on the amount of land under communal stewardship. Nonetheless, field reports indicate that most of the forest land in Ratanakiri Province, and probably Mondulkiri as well, falls under the utilization of indigenous communities who coordinate with neighboring communities; make decisions on forest protection, utilization, and fire control; and attempt to maintain forest cover in the face of growing pressures from outside land speculators.

Figure 1. Zoning of Suggested Forest Tenure Based on Distance from Village Showing Location of Forests' Special Significance to Village Culture

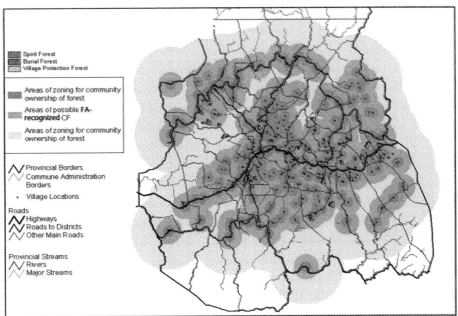

A zoning exercise conducted in Ratanakiri Province in 2005 indicated that much of the province could be zoned for community forestry, assuming all forest within a 2-, 5- and 10-kilometer radius of each village was allocated for communal ownership or placed under CF agreements.[10] Under such a scenario, community stewardship of the province's natural resources would be the primary instrument for managing economic growth based on agro forestry and forest products, combined with cultural and conservation-oriented tourism.

A national assessment conducted in mid-2005 by Community Forestry International (CFI) found that over one million hectares of forest was under active village management belonging to 511 communes.[11] This data indicates that one-third of the nation's commune may have been engaged in various forms of forest management covering 10 percent of the country's forest area, even prior to the initiation of the NCFP. Unfortunately, the process of recognizing CF groups has been constrained by a lack of human and financial resources within the FA. By mid-2011, only 389,021 ha of forest had been reported to the national Community Forestry Office, of which 237,844 ha was under CF agreements.

While the FA has accelerated the process of CF agreement issuance in recent years, the MOE has been slow to formalize a Community Protected Area program that would extend rights of resident communities in national parks and wildlife areas. In contrast, in 2011–2012, the MOE reportedly opened up 23 national parks and wildlife sanctuaries to private sector investment, issuing ELC and exploration rights to 346,000 hectares in 13 protected areas.[12] A 2004 assessment identified 69 community protected areas covering 66,498 hectares though these had not been formally recognized by the MOE.[13] There is little question that by mid-2012, allocation of Cambodia's forests to investment firms rapidly outpaced civil society and RGC programs to devolve and formalize community rights to the same forests. In spite of the RGC policies prioritizing the commercial development of forest lands over community-based management, apparent in rapid ELC issuance, the community forestry movement in Cambodia has made substantial progress over the past two decades.

EARLY HISTORY OF CF SUPPORT PROGRAMS

Throughout the 1990s, community forestry maintained a low profile in Cambodia's forest policy sector. While forest concessions dominated Cambodia's forestry sector throughout the 1990s, community forestry initiatives developed gradually during the same period with far less publicity. Externally aided community forestry projects began emerging soon after Cambodia began receiving international development assistance in the early 1990s. In 1992, the Mennonite Central Committee began supporting pilot projects in Takeo Province, and soon after Concern Worldwide initiated CF field activities in Kampong Chanang and Pursat Province. Gradually other international NGOs and Cambodian NGOs began experimenting with a wide variety of approaches to community forestry. FAO supported the first provincial strategy for community-based natural resources management in Siem Reap in 1995, a program that grew in coverage as it increased its emphasis on participatory fisheries and forest management over the next ten years.

Throughout the 1990s, a number of staff of the Department of Forestry and Wildlife (DF&W) frequently participated in national, regional, and international meetings on CF subjects. By 1994, some planners within the DF&W were already beginning to explore possible policy and legal frameworks for CF. As a result of these efforts, RGC policies and programs that engage rural communities in natural resource management, including the NCFP, build on over 20 years of work by communities, NGOs, researchers, and development agencies that have established and enabled the evolution of community resource management systems after a lengthy and disruptive civil war. By engaging in research, pilot projects, legal rights efforts, livelihood strategies, capacity building, and the creation of new civil society institutions, they have helped create a foundation on which community-based resource management can be formalized and their rights recognized.[14] In some cases, CF activities have been undertaken by remote communities without external assistance; in other situations, local NGOs have worked closely with villages to create awareness, institutions, and activities that catalyze community forest protection. The staff of the Forestry Administration has worked steadily with NGOs and development agencies over the past two decades to create both a legal and policy framework, as well as an operational program for CF.

CIVIL SOCIETY INPUT INTO CF POLICY FORMULATION

Throughout the mid-1990s, interagency meetings were held to share early field experiences with CF among practitioners, and by 1998 national working groups and networks were being established to explore national approaches to community forest management in Cambodia. In 1998, the Department of Forestry and Wildlife (DF&W) established a Community Forestry Unit and began building a national CF database, supported by GTZ's Sustainable Management of Natural Resources Project (SMRP) (GTZ stands for Gesellschaft für Technische Zusammenarbeit, or German Technical Assistance) and later by the Cambodian German Forestry Project (CGFP). As the national timber concessions program declined in 2000, a growing number of development agencies, and international and national NGOs gave increasing attention to developing community forestry strategies that targeted areas with cancelled timber concessions, well-stocked older growth forests, as well as areas with open-access degraded natural forest. The approach involved supporting the formulation of an enabling legal environment that extends legally recognized management rights and responsibilities to rural communities for state public forestland.

A national working group soon evolved into a CF Strategic Planning Team, composed of individuals from DF&W, GTZ, JICA (Japan International Corporation Agency), ADB, Concern Worldwide, IDRC (International Development Research Center), and FAO, and issued the first National Community Forestry Strategic Plan in 2000.[15] With a growing consensus forming around the National CF Strategic Plan, the CF Strategic Planning Team turned its attention to the need for a legal framework to enable CF to take place formally. In 2001, a Task Force was established to develop a draft CF Sub-Decree (*Anukret*) that could be distributed for public comment. A six-month consultation, drafting, and comment period was held to gain input from stakeholders around Cambodia. Upon completion of a final version in February of 2002, the drafting team submitted the document to DFW for internal review prior to submission to the MAFF and ultimately the Council of Ministers.

Over the next 6 months, a continuing dialogue took place between the DF&W and members of the Task Force in an effort to agree on key elements within the CF Sub-Decree. Aside from dealing with such issues as benefit sharing, CF management plan approval procedures, and committee structures, a key goal of the Task Force during the entire process was to ensure that the draft Sub Decree (*Anukret*) was informed by a broad group of stakeholders and that there was a strong sense of ownership for the CF Sub-Decree, both within government and among donors and NGOs.[16] Part of the success of the consultation process was the presence of several NGO networks and forums that facilitated input from their members, together with the ability of government and non-government members of the Task Force to work collaboratively.

For over a year, the Community Forestry Sub-Decree was reviewed and revised by the DF&W as comments were received from key stakeholders, MAFF, and the Council of Ministers. In late 2003, the CF Sub-Decree was finally approved. The DF&W was reorganized in early 2004, adopting a new organizational structure and renamed the Forest Administration (FA). In 2008, the CF guidelines (*Prakas*) were approved and the Community Forest Unit was upgraded to an Office to guide the implementation of the NCFP that was sanctioned in 2010 as part of the National Forestry Program.

While considerable effort went into the development of the national policy and legal framework for CF, a major component of some projects was the provision of technical training and funding to Cambodian NGOs and forest-dependent communities. A number of international NGOs provided support to local NGOs, Buddhist pagodas, and other small civil society and community-based organizations that were emerging and encouraging the expansion of community forestry groups across the country. Such projects often focused on CF training and capacity building, awareness building and CF extension, provision of small grants to forest communities, CF mapping, and CF networks and partnership building. Assistance from GTZ, CFI, JICA, FAO, RECOFTC (Regional Community Forestry Training Center) and other organizations to the FA's Community Forestry Office (CFO) helped to create capacity within local FA and government offices.

A number of organizations and programs contributed to the consensus that emerged between some civil society organizations and RGC staff that a multifaceted CF support strategy in Cambodia was needed to achieve a variety of social and policy impacts, economic benefits, and improvements in natural resource management. The Community-based Natural Resource Management Learning Institute played a strategic role in bringing together key Cambodian and expatriate professionals to identify core concepts and experiences that should guide the country's participatory management strategy.[17] The Cambodian Development Research Institute (CDRI), the World Wide Fund for Nature (WWF), and the Wildlife Conservation Society (WCS) conducted important research activities including those assessing the value of forest products to rural communities. The following sections present key CF support strategies that are being used to facilitate the forest sector transition over the next few decades.

CIVIL SOCIETY COLLABORATION IN SUPPORTING CF

Building community, civil society, and government capacity to implement the NCFP and other CBNRM strategies will be a necessary component for Cambodia's broader forest sector transition to a decentralized mode of operation where communities possess meaningful rights in management decision-making.

In Cambodia, many forest-dependent communities reside in remote areas with little access to mass media or information about government programs, including community forestry. This presents challenges to all stakeholders regarding how best to engage forest-dependent communities with forest management problems and disseminate information regarding RGC laws, policies, and programs that support villages interested in protecting their forests. In building awareness regarding community forestry programs, as well as land and forest rights, livelihood activities, health, and other issues important to rural families, a variety of organizations are playing important and distinctively different roles. One organization delivered information regarding community forestry through traditional Khmer shadow puppet theater.

Activist Buddhist Monks and their institutions are playing a growing role in creating community awareness regarding the need for forest conservation in many parts of Cambodia. As they are held in high esteem in Khmer society, they have considerable credibility and influence in rural communities and have often been effective in mobilizing community forest protection activities. An example from Oddar Meanchey is presented in chapter 6, though there are many other cases throughout the country. The role of Buddhist monks in raising environmental awareness is important as it presents a non-western perspective on the value of conservation. On Cambodia's border, in northeast Thailand, some monks are also playing the role of environmental advocates. One scholar notes that "at the periphery of globalization are the disaggregated peasants and a reformulated politico-religious counter-ideology, articulated by activist monks.[18]

Indigenous Ethnic Groups

A number of organizations have worked closely with indigenous ethnic groups in Cambodia to assist them to address problems of forest loss, erosion of land rights, marginalization, and poverty. Much of this activity has focused on the northeastern parts of Cambodia including Ratanakiri, Mondulkiri, and Kratie.

Community meetings that discuss threats to land, forests, and natural resources, and that develop strategies to address them have helped some communities respond to growing external pressures on their resources.

Mapping community lands and resources, developing sustainable use plans, and building multi-village networks and federations have proven to be effective in strengthening the position of often disempowered indigenous ethnic minorities. An example of this process is presented in chapter 5.

In addition, NGO support organizations have also worked closely with Cambodian legal aid groups to find ways to better support indigenous ethnic groups through the law. The Cambodian Land Law of 2001 grants collective land ownership rights to indigenous communities, as it is recognized that land and forest alienation is one of the major threats to the indigenous peoples in Cambodia. Logging, commercial plantation development, commercial mining, and illegal land appropriation are threatening to destabilize these communities, which make up approximately two-thirds of the population of Ratanakiri and Modulkiri Provinces.

While the assistance from often small NGOs with limited resources has helped raise community awareness, build networks, and increase government and public attention to land and forest conflicts impacting indigenous ethnic minorities, such problems have grown rapidly over the past decade with frequent public protests.[19] The issuance of large ELCs in remote forest tracts often overlap with the lands of indigenous ethnic minorities. Rapid deforestation that follows ELC issuance for rubber and other plantation crops quickly impacts local communities who are generally offered little or no compensation. While protests are growing, many communities are intimidated by powerful people who hold the concession. As one Kuy person said, "Although we are so fed up for being mistreated, we are so scared to protest since we have been under threat."[20]

Community Forestry Groups

Many organizations have played a role in supporting the formation of CF groups for over twenty years, increasingly in collaboration with Forestry Administration staff. This process has helped establish approximately 500 CF groups around the country that have some level of training in forming executive committees, establishing by-laws, organizing community meetings, formulating rules and regulations, and organizing protection and restoration activities. In some cases, CF groups have developed detailed management plans for their forests.

With financial support over the past decade from FAO, Concern, Oxfam G.B., CFI, RECOFTC, and other organizations. Local NGOs based in the provinces have begun acting as effective CF group mobilizers, providing organizational and technical guidance and sometimes functioning as a communication bridge with RGC technical agencies. While this process has been effective in accelerating the formation of CF groups and building their capacity, it has been constrained by erratic funding from donor agencies, uneven support from local FA officers, and lack of ability to effect high-level decisions regarding ELC issuance.

FORESTS AND COMMUNITY LIVELIHOODS

Successful CF strategies need to be based, not only on capable community institutions and support organizations, and on secure tenure and management rights, but also on viable economic activities based on sustainable forestry practices. Fortunately, rural Cambodia communities have considerable experience in utilizing forest resources for subsistence and commercial purposes in a historically sustainable manner. For centuries, NTFPs (non-timber forest products) have been a core component of the economy of Cambodia's forest communities. A long history of trade between Cambodia, Indonesia, and China in forest resins, medicines, canes, exotic foods, and other products is well documented. The Asian rattan trade alone is valued at over $1 billion per year.

While international trade now focuses on manufactured products and major commodities, NTFPs remain an important source of income for millions of rural households, although most NTFPs are collected by women and children and used for subsistence needs. There are considerable opportunities to increase the value of NTFP production systems in Cambodia through enhanced management techniques, improved processing, and better access to markets. NTFP productivity can be increased through better management of the forest ecosystems, which may involve fire control, pruning, and thinning, as well as *in situ* cultivation and improved harvesting techniques. Value-added processing can substantially contribute to increased returns to producers and may include better cleaning, refining, and packaging.

Improving market linkages can increase household income by reducing the number of middlemen involved in bringing a product to market. Market surveys of NTFPs in Oddar Meanchey Province indicate that many NTFP prices in village markets may be 50 percent or more below prices available in district markets (see table 1). An important component in improving NTFP marketing involves connecting gatherers with district and national markets, while also reforming forest policies to remove restrictive transit fees.

At the same time, NTFP support programs that access higher-end markets can also increase returns to producers. In some cases, profits can be further enhanced through product certification. Recent studies indicate the importance of NTFPs to forest communities throughout Cambodia. In Mondulkiri, for example, 82 percent of the population is composed of Phnong and other ethnic minorities who primarily practice shifting cultivation. Another study of forest households in three provinces found that NTFPs are estimated to account for approximately 15 to 20 percent of all household income.[21]

Table 1. Comparative Prices of NTFP at Village and District Markets in Oddar Meanchey Province, Cambodia, 2006[22]

Product	Village Market Price in Riel	District Market Price in Riel
Wood mushroom	500 to 1,000 per kg	1,500 to 2,000 per kg
Banteng mushroom	1,500 to 2,000 per kg	3,000 to 4,500 per kg
Rattan mat	35,000 to 45,000 per piece	50,000 to 61,000
Bamboo shoot	500 to 700 per kg	600 to 1,000 per kg
Sluk pric (Melientha suavis)	1,000 to 8,000 per kg	1,500 to 15,000 per kg
Wild ginger	300 to 500 per kg	500 to 700 per kg
Solid resin	700 to 1,000 per kg	1,500 per kg
Liquid resin	570 to 670 per liter	700 to 840 per liter
Honey	8,000 to 10,000 per liter	11,000 to 12,000 per liter

HONEY COLLECTION

A World Wide Fund for Nature (WWF) project team found that honey collection was the second most important source of income after agriculture for the province's tribal households.[23] Over 52 percent of families reported collecting honey in the dry forest using traditional methods. Each tree may have twenty to thirty hives that produce from 60 to 150 liters of honey per tree. Yet, despite high levels of productivity, profits are low. They are constrained by a number of factors including over-collection, deterioration of the forest habitat, poor market access, adulteration of honey with sugar and water, and the use of inappropriate tools.

RESIN

Resin is one of the most important non-timber forest products. It is estimated that 20,000 tons of resin are collected each year in Cambodia, representing an important source of income and employment for 100,000 households. Market prices range from $172 to $325 per ton.[24] Resin trees which belong to the *Dipterorcarpus* species (*alatus, intricatus, turbinatus, and costatus*) are typically 60 centimeters at chest height, and it is customary for them to "be owned" by the first person to tap them. This customary tenure arrangement, is unique to resin trees, as other trees and NTFP are considered to be accessible to all.

One study found that in Kompong Thom province, an average household "owned" 260 resin trees, versus 100 trees in Preah Vihear, and 80 trees in Mondulkiri. The study also found that the collection area averaged around 13 kilometers from the villages, with trips lasting up to 4 days. Villagers are reported to make 15 to 40 taping trips each year, usually after the rice harvest.[25] This would indicate that the estimates reported earlier in this chapter regarding community use of forest areas within a 10-km radius of the village realistically reflects collection areas, at least for resin.

WOOD AND OTHER COLLECTION

In addition to resin collection, villagers collect many types of forest products. Firewood is a primary source of fuel from approximately 90 percent of the country's households. Timber for house construction, rattan, vines, bamboo, thatch, mushrooms, wild fruit, edible leaves, and medicinal plants are all key materials for rural families. Forest resources also play an important role in times of crisis. Forest foods supplement the family diets during times of scarcity. Collection of NTFPs can also generate badly needed income when households are short of cash as a result of illness, crop losses, droughts and floods, robbery, or death. In Preah Vihear and Kompong Thom, one study found that 62 and 83 percent of households faced food shortages, illnesses, or other family crises, and often relied on forests to supplement incomes.[26]

These findings illustrate the important economic role forests play in the lives of many rural Cambodians. As forests are degraded by logging and resin trees are felled, and as forests are cleared for agro-industrial enterprises, forest-based sources of livelihood decline, eliminating an important source of materials in kind and cash flows that support families in times of crisis. RGC decisions to grant ELCs need to consider the social and economic costs of such actions for rural communities.

CONCLUSION

There is a strong and growing grassroots demand for greater forest and land tenure security among Cambodia's forest-dependent villages, one that appears to be rising in many parts of the country as forests are threatened by ELCs, mining exploration concessions, and other development activities. As the Independent Forest Sector Review noted, "Community forestry has demonstrated that local people are concerned about forests and are prepared to organize around forest protection and management."[27]

The perceived need of communities to secure effective control over local natural resources is often driven by concerns over the extensive land speculation, illegal logging, ELC issuance, and plantation establishment taking place in many areas, often at the expense of community lands. The FA and the NGO community is receiving an expanding stream of requests from rural communities for forestry support, reflecting their growing concerns and perceived need to protect local forests. Given that the resource in question is state public and private forest land, including national parks and wildlife areas, strategies are required to formalize resource rights and responsibilities through the creation of partnerships between the respective villages and the RGC.

NGOs have played a strategic role in facilitating the development of a national policy framework and civil society capacity to implement community forest management in Cambodia. They have pioneered a variety of cost-effective field strategies for transitioning to legally recognized community-based forest management. NGOs have assisted the RGC in bringing together stakeholders to design a national implementation plan, building the capacity of rural villagers and local NGOs to mobilize community forest stewardship, and developing a funding base to reorient forest management in ways that should lead to improved rural livelihoods and better forest conservation. Buddhist monks, an important part of Cambodia's civil society, are also increasingly active in promoting forest conservation and engaging communities in forest protection.

Cambodia is currently one of the poorest nations in Southeast Asia, with limited government resources to support rural development and natural resource management programs. The institutional infrastructure is weak, both in terms of local government as well as the capacity of the Forest Administration at the field level. Most cantonments have few triage staff (field level), especially those with experience in coordinating with commune councils and communities. Commune councils are also new institutions with little experience in resource management and often preoccupied with other priorities including roads, schools, and health clinics. Over time, as field management activities grow, CF committees can be coordinated through Commune Councils. Initially, however, the emphasis will likely be placed on building resource management capacity at the hamlet and village level, with coordination and networking facilitated through Commune Councils and District and Provincial level government.

While there has been growing support for CF in Cambodia, many challenges remain. Little funding exists within the FA and the MOE to support CF programs and activities, and international donor support is based on short project periods, which makes it difficult to achieve long-term goals. The legal and policy environment emerging in Cambodia presents both supportive and constraining elements for CF.

The CF implementation guidelines (*Prakas*) describe in considerable detail the actions rural communities must take to receive formal recognition of their community forest areas. The limitation of the guidelines is that the terms for community participation require organizational capacities that many villages do not yet possess, while the FA retains significant control over resource utilization, complicating the interface with rural communities that may approach forest management differently. Detailed prescriptions, regarding the structure and function of village organizations, and formulation of rules and regulations, may overwhelm isolated villagers who have little familiarity with modern governance and bureaucratic procedures.

As in other countries, Cambodia's CF legislation requires the adoption of a "national" model of CF that presently exists largely as a concept, while actual CF groups take a variety of forms, and often reflect more traditional modes of communal organizations and resource use practices. There are an increasing number of local NGOs that are gaining experience in CF, fostered by small donor grant programs, and they may be in a position to facilitate interactions between FA territorial staff and forest-dependent communities as the national CF program unfolds. The successful implementation of CF laws and the NCFP will likely depend on flexibility on the side of both the government and communities, to find common ground where viable management partnerships can be established. There is also an urgent need for a broad-based effort to build the capacity of forestry staff, NGOs, local government, and community leaders to work together and restore and develop the nation's forest resources.

There appears to be growing concern in many rural Cambodian villages over rampant illegal logging, land encroachment, and commercial plantation expansion. This concern is reflected in local leaders, monks, and NGOs who are attempting to organize forest protection initiatives. Government agencies, NGOs, and local government units need to find ways to assist communities to organize forest protection activities, empowering them through the Land Law, Forest Law, and CF Sub-Decree. While Cambodia's transition to CF as a national strategy is still in an early stage of development, many important elements have been put in place over the past decade. If government and long-term donor support can be mobilized, several million hectares of forest land could be formally brought under community stewardship over the next ten years. This will require a long-term commitment from the donor community to finance the transition in the forestry sector, as the country transfers formal management authority to rural villages.

Development organizations need to continue to encourage and support the Forestry Administration to proceed with the implementation of the National Community Forestry Program and the Community Forestry Sub-Decree in order to reduce the incidence of deforestation and forestland conflicts, conserve biodiversity, and address rural poverty concerns. A parallel effort is also required to build civil society organizations, including local NGOs, academic and training institutions, and community federations and networks. Through an integrated strategy that provides incentives and supportive mechanisms to key stakeholders, it is anticipated that Cambodia's forest sector can continue to transition toward greater community and local government control, resulting in enhanced social and environmental stability. As community forest tenure is secured and capacity developed, improving rural livelihoods needs to be addressed through local forest enterprise activities, including NTFPs and marketing, small-scale timber milling for domestic markets, and ecotourism.

From 2011 to 2012, the rapid issuance of ELCs and mining exploration rights in state forests and protected areas contrasted sharply with a much slower rate of CF agreement issuance. The encroachment of agro-industrial enterprises into forest areas, informally managed and utilized by rural Cambodian communities, has created growing conflicts. There is an urgent need to create awareness within the national government planning and land allocation process that community-based resource management systems represent a viable alternative to ELCs, due to their multiple social, economic equity, political, and environmental benefits.

CASE STUDY
EXPERIENCES WITH HONEY ENTERPRISE DEVELOPMENT[28]

AMANDA BRADLEY

CFI (Community Forestry International) supported the development of a honey enterprise in the mangrove forests of Koh Kong Province from 2006 to 2008. The goal of the project was to develop a pro-poor trade strategy that identified high-value "green" products for niche markets, a strategy promoted by the RGC. The project hoped to develop markets for sustainably harvested forest products to provide livelihoods in biodiversity-rich protected areas.[29] To access international luxury markets, a product must meet the following criteria: quality, quantity, timeliness, price, and "story line." Buyers are willing to pay a premium price for products that are sustainably produced (environmentally friendly) in socially responsible ways (fair trade), and that have health benefits (organic or medicinal).[30]

Cambodian forest honey has the potential to meet all these criteria. In the private sector, several companies are already packaging and marketing Cambodian honey to upscale tourist and private markets. These enterprises, however, are largely buying from middlemen and doing little to improve value chains for the collectors or to improve processing. One exception is the Cambodia Biologicals Company Ltd. that relies on organic and fair trade approaches, and with whom CFI developed a joint effort to develop a community-based honey production and marketing enterprise.

The project was located in Phnom Toub Cheang CF near the town of Sre Ambel in Koh Kong Province. The village is located along the Saom River that flows into the Gulf of Thailand. The community has been managing its community forest since 2003 as a response to the Malaysian Samling timber concession that threatened local forest resources. With support from the American Friends Service Committee (AFSC), the village completed its CF bylaws, rules and regulations, and map and management plan as part of the process to receive recognition from the FA. The community forest covers 3,046 hectares of evergreen and deciduous forests, as well as some stands of pine species and some bamboo and rattan forests in disturbed areas. A dominant tree in the forest is *Melaleuca cajeputi* (*smach deng*), which is an excellent tree for honey bees as its florescence attracts them to the area. Honey collected from these areas has the pleasant scent of the melaleuca flower.

The project team found that all four of Cambodia's major bee species were present in the area. Honey is collected in the early part of the rainy season in lowland areas, but as the *melaleuca* flowers decline in September, the bees move to the mangrove forests along the coast, returning in October. Men are primarily involved in honey collection as it requires climbing trees. There is no ownership of honey trees or collection areas, unlike some other parts of Southeast Asia, where honey trees are owned and rights are strongly held. Women play a dominant role in honey processing and dealing with honey traders. At the beginning of the honey season, honey collectors create rafters in the forests for the bees to construct hives on. Once rafters are built, no one disturbs the hives.

The community responds to market prices. When the price is high, sixty families may participate in honey collection, with each family collecting up to 200 liters during the rainy season at a farm gate price of around $2.50–$5 per liter. If prices are low or supply limited, fewer villagers participate. Under the project, certain criteria were agreed upon to ensure sustainable production, a higher quality product, and improved social and management systems. They included the following:[31]

- Honey collectors must be registered as community forestry members.

- Only 80 percent of the hive "honey head" should be harvested in order to provide food for the bee colony.

- Honey must be harvested in daytime only, to avoid disturbing the colony.

- Containers must be clear and properly labeled (weight, name of collector, location, and date).

- Honey must be pure (not adulterated with water or sugar).

- Honey head cannot be stored more than three days to avoid spoilage.

- Ownership of the rafters must be respected.

- CFMC (Community Forest Management Committees) must be active in honey purchasing.

- Financial management must be transparent and accountable.

The project team provided training and a purchase agreement to receive honey at a fixed rate of 18,900 riel per kilogram. In July 2007, the project received the first shipment of approximately 600 kilograms of honey, which overwhelmed the absorptive capacity of the project. Gradually, the system adjusted, eventually receiving ten honey shipments between July and October 2007 that totaled 1,286 kilograms, and with a total income earned by the 174 families of $7,722.

The project found a number of key issues that impact the development of honey enterprises, many of which also apply to other NTFPs. These include the following: preventing theft of honey from other community members; ensuring honey is of a high quality and not adulterated; establishing a viable collector group with a supportive management structure; stabilizing supply to ensure that marketing commitments can be made; timing the production to reflect seasonal variation; acquiring the technical support needed in the field, such as digital scales; and, finally, sharing benefits in ways that create incentives for individual families and the community as a whole to participate.

The project identified a number of key lessons that are applicable to most NTFP development initiatives, including: working with experienced organizations to streamline honey collection and purchases, collecting and transporting a large enough volume to reflect the costs of delivery, processing and transporting efficiently, paying collectors promptly, making purchase lots traceable to producers to control quality, and providing training to ensure collectors understand buyer requirements.

NOTES

1. Bruce McKenney, Yim Chea, Prom Tola, and Tom Evans, *Focusing on Cambodia's High Value Forests: Livelihoods and Management* (Phnom Penh: Cambodia Development Resource Institute and the Wildlife Conservation Society, 2004).

2. FAO, Forest Resources Assessment 1990: Tropical Countries. Food and Agriculture Organization of the United Nations, Rome cited in A. Mittelman, "Secondary Forests in the Lower Mekong Subregion: An Overview of Their Extent, Roles and Importance," *Journal of Tropical Forest Science* 13, no. 4 (2001): 671–90, 676.

3. A. Mittelman, "Secondary Forests," 677.

4. Ibid., 68.

5. Ibid., 68.

6. Paul Vrieze and Kuch Naren, "Carving Up Cambodia One Concession at a Time," *The Cambodia Daily* (March 10–11, 2012), 7.

7. Jurgen Fichtenau, Ly Chou Beang, Nup Sothea, and Dy Sophy, "Assessment of Ongoing Community Forestry Initiatives in Cambodia" (Phnom Penh: DFoW/GTZ, 2002).

8. McKenney et al., *Focusing on Cambodia's High Value Forests,* 5.

9. CFI eds., "Proceedings of the Non-Timber Forest Product (NTFP) Workshop and Seminar" (Phnom Penh: CFI, 2006), 26.

10. Brown, Graeme and Alistair Stephens, *Mapping of Possible Community Forest Management Tenures in Ratanakiri Province, Cambodia* (Phnom Penh: Community Forestry Alliance for Cambodia (CFAC), forthcoming 2005), 26.

11. Amanda Bradley, Mark Poffenberger, and Ponreay, "National Survey of Community Forestry in Cambodia" (Draft document, Phnom Penh: Community Forestry Alliance for Cambodia, 2005).

12. Vrieze and Naren, "Carving Up Cambodia," 7.

13. Toby Carson and Hou Kalyan, "Overview of the Past and Present Situation of CBNRM in Cambodia," in *Emerging Trends, Challenges and Innovations for CBNRM in Cambodia* (Phnom Penh: CBNRM Learning Institute, 2009), 26.

14. Amanda Bradley and Phuong Pichponreay, "Community Forestry Seed Grants" (Phnom Pehn: CFI, 2006); Graeme Brown, Jeremy Ironside, Mark Poffenberger, and Allistari Stephens, "Forest Stewardship in Ratanakiri: Linking Communities and Government" (Phnom Penh: CFI, 2006).

15. Royal Government of Cambodia, *Final Draft National Community Forestry Strategic Plan* (Phnom Penh: Department of Forests and Wildlife, 2000).

16. RECOFTC, *Balancing Ownerhsip: Overview of Writing Retreat about the Community Forestry Consultation Process in Cambodia* (Phuket, Thailand: RECOFTC Training and Report Workshop Series, 2002).

17. CBNRM Learning Institute, *CBNRM in Cambodia: Selected Papers on Concepts and Experiences,* Vol.1 (Phnom Penh: CBNRM-LI, 2005) and CBNRM Learning Institute, *Emerging Trends, Challenges and Innovations for CBNRM in Cambodia,* Vol. 2 (Phnom Penh: CBNRM Learning Institute, 2009).

18. Jim Taylor, "Thamma-Chaat: Activist Monks and Competing Discourses of Nature and Nation in Northeastern Thailand," in *Seeing the Forest for the Tree,* ed. Philip Hirsch (Chiang Mai: Silkworm Books, 1996).

19. Vrieze and Naren, "Carving Up Cambodia," 11.

20. Ibid., 10.

21. Ibid., 61.

22. CFI eds., "Proceedings of the Non-Timber Forest Product."

23. Ibid.,12.

24. Ibid., 36–40.

25. McKenney et al., *Focusing on Cambodia's High Value Forests,* 58.

26. Ibid., 62.

27. Independent Forest Sector Review, *The Forest Sector in Cambodia: Part I – Policy Choices, Issues and Options* (April 2004), 68.

28. This case study is drawn from Amanda Bradley and Andrew McNaughton, "Community Forestry and Honey Enterprise Development" (Phnom Penh: CFI, 2007).

29. Ibid.

30. Ibid., 7.

31. Ibid., 17.

COMMUNITY REDD+ IN ODDAR
MEANCHEY PROVINCE

MARK POFFENBERGER

THE UNITED NATIONS FRAMEWORK CONVENTION ON CLIMATE
Change (UNFCCC) is exploring a mechanism to reduce emissions from
deforestation and forest degradation (REDD) to address global warming.
The core concept behind REDD is that deforestation trends can be slowed,
halted, or even reversed, conserving billions of tons of carbon that would
otherwise be emitted into the atmosphere and increase global warming.
REDD+ goes beyond addressing only forest degradation and deforestation and includes the role of conservation, the sustainable management of
forests, and the enhancement of forest carbon stocks.

To succeed, REDD+ projects will need to control powerful drivers
of deforestation and forest degradation operating at multiple levels and
carried out by a variety of actors, from rural people to political and
economic elites. This case study of a REDD+ pilot project in northwest
Cambodia explores how drivers might be contained by forest communities working in cooperation with the Forestry Administration (FA) and
civil society organizations. If successful, REDD+ strategies at the national
and subnational level have potential to generate substantial revenue to
support community-based forest stewardship through the sale of carbon
credits on international voluntary and compliance markets.

REDD+ strategies have potential, if structured appropriately, to enable and finance a broader historic forest transition in the developing world. In Asia, over the past twenty years, a new generation of forest policies has been developed after a century of industrial logging, which resulted in the degradation of tens of millions of hectares of once dense tropical forests. Community-based forestry has the potential to restore watersheds, conserve biodiversity, improve local livelihoods, and sequester carbon, but funding to support these sustainable forest management transitions has been erratic and inadequate. Further, international policies and markets supporting REDD+ initiatives have been slow to develop. As a result, some of Cambodia's poorest people, dwelling in remote upland watersheds, are struggling to protect and restore local forests with few external resources. REDD+ programs could create a framework for financing community-based forest conservation on a national scale and long-term basis, while establishing empirical methods to quantify and value important environmental services.

CAMBODIA AND REDD+

Cambodia provides an interesting context to explore the feasibility of REDD+ project development at a national and local level. After nearly three decades of political instability, the national social and economic infrastructure was extremely weak, and economic growth was a priority for government policymakers and donor agencies. Timber exploitation was seen as an answer and expanded rapidly during the 1990s. During this period, approximately 60 percent of the country's forests were leased to private concessions. Industrial forestry proved to be a failed strategy for the country, with unsustainable exploitation leading to widespread forest degradation, while generating limited income for the national government. In 1997, an external review team estimated that the state collected only 12 percent of the revenues due from logging, while actual extraction was nearly ten times the sustainable yield.[1]

Since 2001, the FA has cancelled or suspended six million hectares of logging concessions and is exploring REDD+ strategies at the national and subnational level as a means of generating revenues to support the National Community Forestry Program (NCFP). REDD+ creates a potential management alliance between an important national government agency and hundreds of forest communities. Whether REDD+ revenues can provide sufficient income to meet national economic development goals remains an important issue. Other uses of the forest, such as conversion to plantation crops under the economic land concession program, might generate a larger income stream, but would frustrate Forestry Administration efforts to achieve the national target of retaining 60 percent of land area under forest cover.

At a local level, REDD+ project development has potential due to the presence of social and ecological capital reflected by motivated communities eager to conserve threatened forests, as well as some of the last remaining lowland evergreen and dry deciduous forest in the province. While the project area has characteristics conducive for project development, it also faces serious challenges. In the late 1990s, after twenty-five years of political insecurity, peace came to Oddar Meanchey Province located in a remote corner of Northwest Cambodia. This sparsely populated, heavily forested region began drawing a growing flood of migrants from the Lower Mekong region, as well as from neighboring provinces. Over the next decade, the population expanded at a remarkable rate of 12 percent per year, while forests disappeared at over 2 percent annually.

With a national policy mandate to preserve 60 percent of the country under forest cover, an alliance between the FA and local communities, financed through REDD+ carbon payments, could be a viable strategy to address the rapid loss of forest cover. While the FA possesses considerable legal rights under the Forest Law to manage much of the province's land area, it inevitably must compete with other government agencies that desire to control these resources. These include the military, the Ministry of Interior, and other units within MAFF. At the local level, forest fires, illegal logging, migrant land clearing, and other activities are also key drivers of deforestation and can only be addressed by resident communities with the support of government agencies and civil society organizations.

The case study on Oddar Meanchey is based on information generated from extensive discussions with local and national stakeholders during the early development and design of the project. The drivers of deforestation and the mitigation activities were identified through a series of social appraisals conducted in 2008 and 2009 that involved forest-dependent communities, local government, FA field staff, and other stakeholders. CFI staff visited most of the thirteen participating forest blocks and villages, and conducted in-depth interviews with forest-dependent families and other stakeholders, which together provided insights into deforestation drivers and possible mitigation strategies. Discussions with the Buddhist monks of Samraong Pagoda and the staff of the Children's Development Association (CDA), both of whom have been instrumental in supporting nascent community forestry management committees (CFMCs) in the province for the past five years, provided additional historical perspectives on the evolution of CF groups and deforestation issues. Information presented in the case study also represents field observations and secondary sources.

PROJECT IDENTIFICATION

In 2007, the 67,783-hectare Oddar Meanchey Community Forestry REDD Project was identified by CFI, a nonprofit organization that had supported the national transition to community-based forest management in Cambodia since 2003. In November 2007, the FA, supported by the joint donor Technical Working Group on Forests and the Environment (TWG-F&E), unanimously approved the REDD pilot project in Oddar Meanchey. With the endorsement of the Cambodian government, a one-year grant from the Danish government, and technical guidance from Terra Global Capital (TGC), a San Francisco-based carbon development company, the project sought to test emerging REDD policies with fifty-eight villages organized into thirteen CF groups in northwest Cambodia.

The project methodology was designed to comply with emerging REDD+ guidelines developed by the Voluntary Carbon Standard (VCS) and the Climate and Community and Biodiversity Alliance (CCBA). The project is expected to sequester 7.1 million tons of CO_2 over the next thirty years and mitigate the impact of a number of drivers of deforestation and forest degradation, while responding to the economic needs of the low-income rural populations that inhabit the project area. Over the past four years, the RGC has formally recognized the CFMCs in the project area under the National Community Forestry Program, providing them with a fifteen-year renewable lease and a minimum 50 percent share of carbon revenues. CF agreements were signed with eleven of the CF groups in May 2009, with the remaining two CF groups legalized in April 2011.[2] The project intends to create a thirty-year income stream that will enhance livelihoods and build natural resource management capacity among the project's 10,000 participating households. In 2009, PACT, a US-based NGO, assumed responsibility for supporting project implementation from CFI.

PROJECT LOCATION

The project is located in northwest Cambodia's Oddar Meanchey Province, which has experienced some of the most rapid deforestation in that country over the past decade, at an average rate of 2.1 percent. Oddar Meanchey is an appropriate setting for a REDD+ project because of its historic high rate of deforestation, and multiple drivers of deforestation and forest degradation, creating a clear need for REDD+ mitigation activities. The project's goal is to slow these trends by supporting communities in the conservation of the remaining forest "islands" and fragments, while also restoring the recently degraded forests.

The community forestry groups in this remote corner of Cambodia have been demonstrating a capacity to protect forests since they began organizing a decade ago. They are motivated to undertake forest protection, actively participating in forest patrolling, not by the illusory promise of carbon payments, but by the realities of their daily lives. Sokh Smit, a forty-one-year-old CFMC member notes, "The people with money in our village go to Samraong Town when they are sick; the poor people go to the forest to collect traditional medicine." Not only medicinal herbs, but much of the daily requirements of rural Cambodians comes from the forest. The timber to build houses, barns, and tools; fibers for basket making; foods ranging from wild fruits and berries, mushrooms, edible leaves, and tubers which are a stable food between harvests—all these make the forest an essential part of the village economy. Equally important are the water sources that healthy, intact forests recharge and protect.

Forest communities like those in Oddar Meanchey are also just beginning to see opportunities in conserving forests to obtain carbon payments. As one women CF group leader said expectantly, "We are going to sell the air to the people who are polluting in the city." The project communities have allies in their efforts to protect threatened forests. Venerable Bun Saluth, an energetic young Buddhist monk who heads the Samraong Pagoda, has mobilized his fellow monks and neighboring villages to protect over 18,000 hectares of forest that still retain leopards, monkeys, alligators, and bears, as well as endangered bird and plant species.

In addition, a local NGO and the FA have assisted the 58 participating communities to form CFMCs and proceed through the steps that lead to a community forestry management agreement under the Community Forestry Sub-Decree. As of 2011, all thirteen participating CFMCs had completed this process and been granted fifteen-year stewardship rights. While many of the project communities are composed of recent migrants who have settled in the area over the past ten to fifteen years, they are motivated to retain their local forests, have substantial local knowledge of the forest ecosystems, and are economically dependent upon forest resources. Due to their physical presence in and around the forests, they are positioned to defend forest resources from illegal logging and further clearing by more recent migrants. According to the chief of Andoung Bor CF, "After signing the CFA [Community Forest Agreement], we have the power. Even if the FA does not come to cooperate we can prevent illegal logging activities."[3] At the same time, they lack the political influence to address more powerful drivers, such as economic land concessions and military encroachment. Their association with the national FA is an essential element in protecting their community forests against such national drivers of deforestation as ELC issuance.

Oddar Meanchey was selected for a REDD project because it possessed characteristics that were conducive for project development, including rapid forest loss, sufficiently large amounts of good forest where deforestation could be avoided, and a national government willing to provide the policy and political support required to address tenure and enforcement issues on public forest lands. At the same time, the province had the social capital needed to contain local drivers of deforestation in the form of highly motivated rural communities, local NGOs, and civil society groups ready to protect forest assets on the ground and restore degraded ecosystems. Finally, the area had ecological capital reflected in both well-stocked forests and recently disturbed forests that showed signs of biological resilience, characterized by vigorously coppicing tree species, high seasonal rainfall, and relatively fertile soils.

PROJECT STRATEGY

The project includes thirteen forest blocks ranging in size from 18,262 hectares to 383 hectares, which are scattered across the central section of Oddar Meanchey Province. The project was designed to include all existing CFs in the province.

Table 1. Project Sites and Communities

ID	CF group name	Commune	District	No. of Villages	Number of households	CF membership percentage*	CF size (ha)
1	Andoung Bor	Koksos	Banteay Ampil	4	746	91%	6,114
2	Chhouk Meas	Koun Kriel	Samraong	1	166	100%	383
3	Dung Beng	Koksos	Banteay Ampil	4	558	85%	1,843
4	Ou Yeay Kaov	Koun Kriel	Samraong	1	177	88%	960
5	Phaav	Phaav	Trapong Prasat	4	429	100%	2,025
6	Prey Srorng	Lumtong	Anlong Veng	5	662	71%	6,344
7	Prey Srors	Kouk Khpuos	Bantaey Ampil	2	246	97%	1,604
8	Ratanak Ruka	Samrong & Koun Kriel	Samraong	16	3,072	76%	12,733
9	Rolus Thom	Koun Kriel	Samraong	4	906	n/a	6,376
10	Romdoul Veasna	Bansay Rak	Samraong	4	878	88%	6,007
11	Samaky	Trapeang Tav	Anlong Veng	4	686	75%	1,079
12	Sangkrous Preychheu	Anlong Veng	Anlong Veng	3	633	82%	4,151
13	Sorng Roka Vorn	Koun Krail	Samraong	6	877	100%	18,164
TOTAL				58	10,036	88%	67,783

*Percentage of households that are members of the community forestry management committee (CFMC)
†Coordinates are in a UTM48N projection with WGS1984 datum

At the same time, some of the initial project area was lost to ELCs and other purposes over the past five years. The project technical support team, currently provided through Pact, an international NGO, has continued to monitor CF activities in the area and engage interest groups within the project structure. CF groups are responsible for the protection of all forests within the project area. Scattered patches of degraded forest in specifically demarcated areas with between 10 percent to 20 percent crown cover will be improved using assisted natural regeneration techniques. The project covers approximately 31 percent of the province's total forest area.

The principle project strategy attempts to build the capacity of local villages to serve as primary managers of REDD+ project forests through a process of education, training, and capacity building. In addition, the project seeks to create a strong coalition of stakeholders committed to achieving the project goals.[4]

PROJECT DURATION AND ACTIVITIES

The project's lifetime is thirty years, excluding the twelve-month project preparation period that involves stakeholder consultations, participatory rural appraisals (PRAs), mapping, boundary demarcation, community training and initial livelihood activities, and negotiations with the RGC, brokers, and buyers. The first five years of the project (i.e., years 1–5) represent the project establishment period, during which time implementation will include the following activities:

- Establish project boundaries
- Control drivers of deforestation and degradation in the project areas
- Develop community project management institutions
- Build REDD+ and A/R (afforestation/reforestation) project development and management capacity in the FA
- Regenerate degraded forestlands within the project boundaries
- Institute monitoring and measurement systems for carbon accounting, biodiversity, and livelihood generation

During years 6–30, the project will move into the maintenance period, and management responsibilities will be supported by the project communities, the FA, and local NGOs. Net revenues from carbon payments during this period will be used to benefit local communities, enhancing livelihoods and improving the quality of the forest. The project start date is July 1, 2008, ending on June 30, 2043. The accounting period and the project period are contiguous.

Figure 1. Map of Oddar Meanchey Province with the Community Forestry Sites

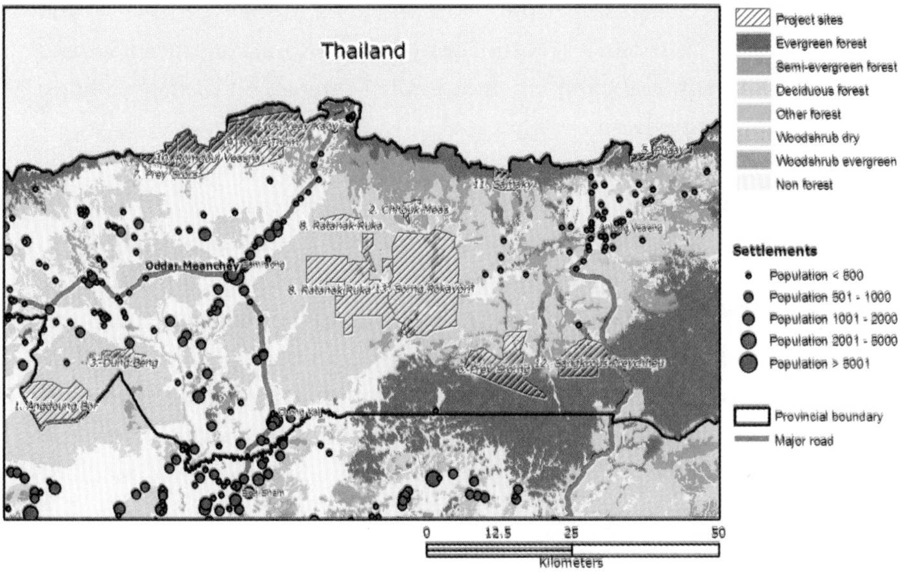

The project lifetime was designed to allow sufficient time to:

- Stabilize and conserve threatened forest cover;

- Restore degraded forests; and

- Build enduring community forest management institutions that encourage livelihood activities supporting the long-term conservation of the area.

DEFORESTATION DRIVERS AND REDD+ MITIGATION STRATEGIES

In the following section, a number of drivers of deforestation and forest degradation will be described (see table 2), including a review of mitigation strategies being proposed by project communities, the FA, and the project design team. The drivers are broadly divided among those that operate at an international, national, and local level. International and national drivers typically require high-level political interventions to mitigate their impact, while local drivers can often be addressed by the community itself, sometimes with assistance from local NGOs and FA field staff. Without a coordinated effort by local communities and national planners, the complex, multi-tiered interaction of deforestation drivers cannot be effectively addressed.

INTERNATIONAL LEVEL

At the international level, commodity prices, investment capital, and high rates of regional economic growth can act as drivers impacting the effectiveness of local REDD projects, as can trans-boundary political and military conflicts. In Oddar Meanchey, for example, it is apparent that the issuance of ELCs in 2007 to Cambodian and Thai private sector interests, driven by global prospects for sugar-based ethanol, accelerated forest clearing. These types of drivers of deforestation are extremely difficult to control at the project level.

Table 2. Drivers of Deforestation and Mitigation Strategies

MAJOR DRIVERS OF DEFORESTATION	POTENTIAL CF MITIGATION STRATEGIES
INTERNATIONAL	
Commodity Markets–rapidly rising prices of sugar, rubber, and palm oil	Controlling commodity prices is beyond the national and subnational project capacity.
Investment Capital–for commercial plantations and land speculation	Transboundary capital flowing into forestland development may be subject to national government control, especially related to policies on issuing concessions to foreign firms or investors.

NATIONAL	
Military–military bases and roads for legitimate defense purposes, as well as support to illegal logging and encroachment on forests by soldiers	National defense needs will likely trump forest conservation, though more routine military demands on forests may be negotiated at the national level—a dialogue between forestry officials and military commanders may be required to resolve illegal activities.
Government Officials–local government officials engaged in illegal land sales and forest clearing	Transparent and public meetings between national government planners and local government officials can communicate the importance of protecting project areas from manipulation and illegal activities.
Economic Land Concessions–large tracts of forestland allocated to private sector firms displace local residents and stimulate social conflict	Senior forestry staff and national REDD project directors need to be in close dialogue with key ministries and committees involved in issuing economic land concessions, as well as long-term public land planning processes.
SUBNATIONAL	
Forest Fires–suppress natural regeneration of degraded forests, create carbon emissions from burning	Hunters, gatherers, farmers and other forest users who often start fires need to be advised and monitored by the community. Fire control strategies require funding, tools, and capacity building to maintain firelines, and suppress fires.
Migrant Encroachment–migrants seeking forestland to farm or resell	Educate migrants regarding community protected forest territory, combined with patrolling, demarcation of boundaries, and sanctions for land clearing.
Land Speculation–forests are felled to establish a claim on land that is later sold, or resold, as land prices increase	Identify middlemen financing forestland grabs and report forest crimes to the police, local government, and forestry agency. Monitor areas. Patrols, boundary demarcation, and signage are also required throughout the project area.
Agricultural Expansion–population growth drives additional forest clearing for agricultural land creation	Community-based land use planning exercises resources and future needs developing plan, maps, and implementation strategy. Design and implement sustainable agricultural intensification project to raise productivity.
Illegal Logging–"high grading" of luxury woods causes ongoing forest degradation and loss of biomass	Limiting access of illegal loggers with small tractors through patrols, trenching along boundary access points, identifying agents and gaining support of forestry agency, police, and military.

Firewood Consumption–90% fuel use from wood with increasing demands from subsistence and commercial users	Introduction of fuel-efficient wood stoves in early project phase, with gradual transition to liquid petroleum gas and solar.

NATIONAL LEVEL

National drivers may be reflected in forest allocation and management policies or related laws and decisions taken by the ministries of agriculture, finance, and the interior, as well as by the armed forces. National market conditions and the state of the political economy are also important forces that can impact forest cover. Subnational REDD projects face serious difficulties mitigating such drivers; however, if linked to a broader national REDD initiative, local projects may reduce their exposure to these forces. Several examples of national drivers and mitigation activities are provided below, drawing on experiences from Oddar Meanchey:

Military Personnel. Oddar Meanchey has been contested domain since the mid-1970s, when it was first occupied by the Khmer Rouge and continually used as a guerilla base until the late 1990s.[5] More recently, tensions with neighboring Thailand have made the province a sensitive area and, consequently, the presence of the military is widespread. Senior officials of the FA's Siem Reap Inspectorate reported over twenty cases of military encroachment into CF areas in 2009. Since defense strategies are classified, it is difficult to get information from the military regarding their operations in the area. In some cases, soldiers are involved in illegal logging, and forest clearing for agriculture and land sales, while in other situations their presence is part of an effort to establish approved "camps" for national defense. According to FA senior staff, each CF area has a unique set of issues and social hierarchy that requires a site-specific approach. A senior FA officer notes that:

> The FA's field staff lives and interacts with the military and this may prevent their ability to confront the soldiers due to their friendships and existing local relationships. Also, the local FA officials may not have the status to gain the attention of senior military officials. The level of FA response needs to reflect the status of the actor.[6]

Mitigation Strategy. Senior officials in the FA have consulted military commanders operating in Oddar Meanchey regarding the community-based REDD+ project and the need to respect these areas and control their soldiers. A joint follow-up visit by other senior FA officials and a military unit successfully evicted soldiers illegally occupying CF areas. The commitment of the FA to resolve this issue is necessary in order to raise the awareness among military commanders. The FA has the authority to defend the integrity of CF carbon project areas due to the endorsement of the project by the Office of the Prime Minister. However, potential future conflicts with Thailand could give the military priority over project areas, including building roads and establishing military camps.

Local Government Officials. In a number of CF areas, commune chiefs and district and provincial government officials are reported to be involved in encroaching on forests within the CF areas. In the past, local government officials had some authority to legitimize forest occupancy, though such letters and documents were often reversed by judicial courts and higher levels of government. Community forestry groups are reluctant to challenge their authority and often need to appeal to higher-level officials in order to control encroachments that have the endorsement of local government officials. Judicial systems are weak and subject to corruption, and they cannot be expected to efficiently address all cases of illicit activities involving local government officers supporting land encroachment. As a result, some mitigation measures can only be taken at the national level.

Mitigation Strategy. A recent national policy limiting the authority of provincial and district governors to grant land concessions has constrained their legal capacity to encroach on REDD+ project areas. At the same time, CFMCs are being supported to actively patrol forest boundaries and install boundary posts and signage to reinforce their claim to project areas. A series of workshops at the district and provincial levels are being organized to inform government officials about the national government's recognition of the CF areas and their inclusion in the REDD+ project.

Economic Land Concessions (ELCs). In recent years, ELCs issued at the national level have been a major driver of deforestation in Oddar Meanchey. Recent ELCs seek long-term leases to convert forestland to plantation crops, including sugarcane and rubber. In 2007, 44,000 hectares of forestland were granted to large concessionaires in Oddar Meanchey, representing 7 percent of the provinces' total land area. ELC issuance frequently results in very rapid and extensive deforestation of the concession area, while displacing forest-dependent populations and transferring pressures that result in degradation or deforestation to neighboring areas. In the case of Oddar Meanchey, ELC applicants and CFMCs were in direct competition for the control of some of the proposed forest blocks in the REDD+ project area. The FA sought the support of the prime minister to secure the CFMC forest areas as a REDD+ pilot project. Without this support, it is unclear if the project would have had the political backing needed to secure the nearly 70,000-hectare project area under community management.

Mitigation Strategy. Mitigating the impact of ELCs on deforestation is certainly beyond the capacity of communities and REDD+ project design teams. ELC issuance takes place at senior levels of government, often with the backing of powerful political and economic actors. In the case of the Oddar Meanchey REDD+ project, with the support of the Office of the Prime Minister and the director general of the FA, discussions where held with MAFF to ensure that ELCs did not conflict with the project area.

SUBNATIONAL LEVEL

Subnational drivers may operate at a provincial, district, commune, or hamlet level and be influenced by a variety of more local actors. National mitigation strategies may have limited impact on forces operating in the field, requiring the engagement of community partners and other local stakeholders. Examples from Oddar Meanchey illustrate some of the forms that local drivers may take and the types of activities being designed to control them.

Forest Fires. Forest fires are natural events in dry deciduous forests in Oddar Meanchey Province. The frequency of fires, however, is greatly increased by human activity. Perhaps 90 percent of dry season forest fires are caused by people, including hunters, children, careless smokers, and farmers burning agricultural residue. In the case of degraded dry deciduous forests, natural regeneration is suppressed due to almost annual ground fires that destroy or damage coppice shoots and saplings. As a result, biomass is lost and regrowth is slowed. At the same time, fuelwood and timber are extracted, leading to a gradual erosion of vegetative material and forest health.

Mitigation Strategies. A number of community members noted that the key to restoring local forests was fire control. If fire can be stopped for four to five years, many young regenerating trees can develop the height and bark thickness to withstand future ground fires. This would indicate that assisted natural regeneration (ANR) strategies implemented by the CFMCs should focus on halting fire in degraded forests that have "high potential" for rapid regrowth. Project activities may include identifying degraded forest sites that have a high density of coppice shoots and saplings, characterized by at least 250 to 300 shoots per hectare, and with good soil conditions and moisture levels. The project participants suggest beginning with areas that are close to communities. The project would also train community members in fire prevention and control techniques, providing equipment and hiring unemployed village youth to act as fire watchers. Project funds would support the construction and maintenance of fire lines with at least five-meter widths.

Migrant Encroachment. Migration into Oddar Meanchey has been very rapid over the past decade. The increase of the rural population is estimated to have been over 12 percent per annum between 1998 and 2008, of which around 9 percent is due to in-migration. Migrants from the Lower Mekong provinces, where land is scarce, hear by word of mouth that the frontier forests of the northwest offer opportunities to secure land. Usually, migrants follow other family members who have already established a "base" in the province. Migrant clearing of small parcels of forest has accelerated deforestation throughout the province and is a major driver of deforestation, though it may be slowing as an increasing amount of the frontier (unoccupied) forest is claimed by local villages, ELCs, and the military. A number of villages noted that migrants were no longer welcome in their communities, and that even local families have no "free" land to expand their agricultural lands. Some national policies may also increase migration. The military may encourage retired soldiers to settle in less densely populated provinces, while the Ministry of Interior may view them as suitable sites for social concessions for landless households.

Mitigation Strategy. Where migrants have settled and communities can accommodate them, they need to be included in CFMC activities and participate in land use planning activities. Where no available forest or unused land exists, communities need to inform migrants of the situation and encourage them to communicate the information to their home villages so that in-migration decreases in that area. Interagency coordination at the national level is also required to ensure that REDD+ projects are not in conflict with policies and programs of other sectors of government.

Land Speculation. Land speculation is often driven by businessmen, officials, and local villagers who seek to claim and clear forestland for sale. According to reports, "power men" hire migrants or local villagers to fell and burn forest cover for $50 to $100 per hectare. Small huts are constructed to indicate residence, although these are frequently unoccupied. This pattern reflects attempts to claim available public forestland and hold it until it can be resold at a higher price. While these actions are usually illegal, letters from local officials are used to create an appearance of legitimacy. Once one plot is opened, it may encourage others to open forests in neighboring areas. In some cases, poor migrants may occupy the plot for one to two years waiting for land values to rise. These plots are then sold to a "consolidator" or businessman who buys a number of them to form a larger plot. Land speculation is slowing as the frontier forests disappear and are claimed by legitimate stakeholders. The global economic recession rapidly cooled local land markets during 2009 and, with fewer buyers, forest clearing for sale has slowed. Yet, this is likely a temporary phenomenon given the rapid population growth.

Mitigation Strategy. Community-based land use planning and forestland demarcation will be accelerated to clarify tenure over remaining forestland in the province, especially in and around the REDD+ project area. Where possible, the project seeks to extend CF areas to neighboring forest tracts and communities, establishing social fencing activities such as community patrolling and facilitating natural regeneration. Meetings and workshops with local government officials and village leaders are essential in controlling illegal land sales. NGO monitoring and reporting of land conflicts and illegal sales to local government officials can make such activities more transparent. Enforcement is also needed to impose sanctions against local officials who organize, facilitate, or provide "official" endorsements of such sales.

Agricultural Expansion. With the province's rural population expanding at an estimated rate of 2,500 to 3,000 households per year, demand for farmland may require an additional 5,000 to 6,000 hectares annually, based on two hectares for each household. Aside from migrant pressures on forest clearing, local communities also require an increasing amount of farmland as children marry and establish independent farms. In the past, village elders were responsible for identifying forest areas that were suitable for rainfed padi fields. Young families or migrants requiring land approached the elders to request a farmland allocation. Usually a two-hectare plot with good soil moisture was selected within the forest. For the most part, these were created before the formation of the CFMCs, and it is difficult for the poor farmers that own them to be evicted. Many CFMCs are accepting these inholding plots if the occupant agrees not to expand the existing fields.

Mitigation Strategy. Interviews with communities in Oddar Meanchey indicate that rainfed rice yields are averaging between 1 and 1.5 metric tons per hectare, while Thai and Vietnamese farmers may obtain yields four to five times this much for the same area. Rapid deforestation in Cambodia has expanded potential agricultural land to approximately 3.5 million hectares. Cambodia's need is not to clear more forests for farming, but rather to intensify agriculture on existing farm lands. Improved farming systems, as well as better access to irrigation, financing, and markets, could allow existing farmland to become two to three times as productive. The project seeks to finance the development of water resources, organic farming, and the introduction of new plant varieties, while improving access to markets. Key components of the strategy include increasing crop yields, shifting to higher-value crops, and raising cropping intensity. Community-based natural resource management plans also need to consider long-term needs for agricultural and settlement expansion around the REDD project site.

Illegal Logging. Illegal logging remains an important driver of forest degradation. Due to the presence of high-value "luxury" timber that is popular for furniture and buildings in cities and tourist areas, timber smuggling is profitable and widespread. Illegal logging results in the "high grading" of the forest, causing degradation, loss of biodiversity, and the creation of access paths and roads—all of which contribute to the further erosion of forest cover and carbon loss. Armed soldiers are often involved in these activities, presenting control problems for weaker community forestry committees.

Mitigation Strategies. Regular CFMC patrols and forest monitoring systems are essential components in creating a deterrent for small groups of illegal loggers. One CFMC reported that in 2008 it had been successful in reducing illegal logging from an estimated 100 cubic meters per year to 20–30 cubic meters. Reducing illegal logging sometimes requires the support of local military commanders, who may be able to restrict soldiers engaged in these activities. The project relies on the FA to facilitate cooperative action between the military and the CFMCs.

Fuelwood Consumption. In Cambodia, 95 percent of the population is dependent on fuelwood for cooking.[7] A study in the neighboring province of Kompong Thom found that the per capita consumption rate was approximately 200 kilograms of greenwood per year.[8] A typical household might consume between 1 and 2 metric tons of fuelwood annually, reflecting 10,000 to 20,000 metric tons of fuelwood used in the project area each year. Some households also burn biomass to create smoke that inhibits mosquito attacks on livestock. Depending on extraction levels and harvesting practices, fuelwood consumption can be an important contributor to gradual forest degradation. This is exacerbated in areas where charcoal production is a local industry linked to urban centers.

Mitigation Strategies. Fuel-efficient stoves are probably the most immediate and cost-effective approach to reducing fuel consumption. Depending on stove type and use practices, this type of technology can decrease wood use by 25 percent to 50 percent. Some stoves also include piped smokestacks that can reduce pollution inside the home and improve family health. Over the longer term, the introduction of liquefied petroleum gas distribution centers and increased household income can reduce dependency on fuelwood. Mosquito control combined with low-cost mosquito nets can eliminate the need for burning biomass to protect livestock.

CONCLUSION

Emerging experience with REDD project development in Cambodia suggests that linking national CF and REDD initiatives with grassroots forest protection activities may be an effective strategy to control multi-tiered drivers of deforestation and forest degradation. This "hybrid" REDD approach is demonstrating that national policy and sector coordination can shelter local REDD projects from more powerful forces causing forest loss. At the same time, in Oddar Meanchey, only through the engagement of forest-dependent people can local drivers be contained.

To date, the Oddar Meanchey Project has received an extraordinary level of political support from the Office of the Prime Minister, based on the strong endorsement of the FA.[9] This likely reflects the desire of the government to (1) conserve a substantial proportion of forest cover, (2) respond to the needs of forest-dependent communities, and (3) generate income from forestland. The REDD+ project framework offers a mechanism to address the first two goals, while still achieving the third element of economic revenue flows.

The RGC views the Oddar Meanchey REDD project as a "test case" to see if payments for forest carbon are a viable alternative to other production-oriented forestland management strategies. In this paper, the discussion focuses on the challenges confronting project managers and participants in controlling the drivers of deforestation. To be successful, the project will also have to show that it is financially viable and that the transaction and implementation costs can be met either through carbon sales or REDD+ funds. Since forest markets are nascent and volatile, and international discussions regarding possible REDD+ funds are still underway, this will require a number of years to assess.

Over the past five years, since the project was identified, a number of serious constraints have emerged, many of them related to problems with new REDD+ methodologies, certification procedures, institutional arrangements, financing, and markets. Early financing for the project during the design and early implementation stage has been limited. After an initial 12-month grant of $350,000 from the Danish government to support community capacity building and forest activities, very little financial support has been received by the 13 CF groups, monks, and support NGO to fund their forest patrols and restoration activities.

The importance of leadership and even modest financial support is apparent in the performance among the 13 CF groups in the project. A recent project case study notes that "Of the 13 CF groups struggling to enforce forest laws with very limited resources, the Monks CF group (Sorng Rokavorn CF) is exceptional in that it has been able to divert donations to the pagoda into forest protection... Sorng Rokavorn CF is the largest and best protected of all the CFs in the province."[10]

Delays in directing funding to CF groups is due, in part, to long delays in securing VCS/CCBA (Verified Carbon Standard/The Climate, Community & Biodiversity Alliance) certification and finalizing carbon sales. Validation of the project is expected in the fall of 2012, five years after design work began. The Voluntary Carbon Standard took nearly two years to approve the project's REDD+ methodology developed by Terra Global Capital (TGC). Problems with the initial forest inventory of 180 plots necessitated a re-survey, further delaying the project. Negotiations between the RGC and TGC over compensation for technical and brokerage support also slowed the design of the project. As one project developer notes, "in the case of Oddar Meanchey, significant emission reductions and revenues are foreseen, but the lack of up-front cash means communities all over the province are struggling to keep up patrolling efforts and demarcate the CF boundaries, putting the project's future benefits in jeopardy."[11]

Lack of investments in community-based, ground level forest protection and restoration undermines the potential of the project to generate REDD+ credits, and while considerable activity is occurring around REDD+ at the international level, resources for innovative pilot projects like Oddar Meanchey remain extremely limited. Over the past five years this REDD+ experience has demonstrated that the global structure for REDD+ to help facilitate project approvals and provide financing has failed to deliver essential support, while at the national and local levels the project is moving forward. The development of the project very likely accelerated the issuance of CF agreements covering nearly one-third of Oddar Meanchey's forest area, arguably the most rapid and extensive series of CF agreements made by the FA up to 2012. Communities, with the support of local monks and NGOs, have continued to struggle effectively in many areas against powerful local drivers of deforestation.

The project would likely not have been possible without the presence of the NCFP (National Community Forestry Program), which provides an enabling policy framework that allows for the legal empowerment of forest communities as resident managers. Over the past decade, Cambodian foresters, academics, NGOs, and development agencies have created this strategy to facilitate a longer-term sector transition that enables a broader engagement of rural villages in the stewardship of the public forest estate. With the closure of large timber concessions that operated unsustainably during the 1990s, large tracts of forest require new management systems, creating a window of opportunity for CF. Prior to the initiation of REDD+, there was limited funding available to support CF groups or technical support services provided by the FA and NGOs. Even when donor support was forthcoming, it was typically site-specific and time-bound, rarely extending beyond a five-year grant.

REDD+ represents a significant shift in the way CF may be funded at the national and local levels, as well as the goals of forest management. A criticism of many national CF and joint forest management strategies is that while communities have received varying degrees of legal recognition, many have experienced limited economic benefits. In part, this results from a lack of adequate funding flowing into CF groups, as well as limitations placed on their capacity to make important management decisions. REDD+ could respond to this problem by creating a thirty-year funding flow that finances CF conservation activities and livelihood enterprises, and injects capital into cash-poor rural communities. This will require equitable and transparent financial mechanisms, with funding based on performance.

Conventional CF projects are often designed by forestry agencies reflecting traditional silvicultural practices. REDD+, with its emphasis on conservation, restoration, and sustainable management, shifts the goals towards carbon storage and sequestration. Conserving forests is often a major goal of forest-dependent communities due to their reliance on NTFPs, water resources, and other services, rather than just timber harvesting. As a consequence, REDD+ project goals and community objectives may be better aligned in comparison to conventional CF projects, which often focus on timber production. REDD+ provides a performance-based framework that creates incentives and allows communities to monitor their progress.

Long-term carbon contracts with buyers, governments, and communities as cosignatories help provide extended tenure security for forest-dependent villages and for forest conservation. This creates a clear opportunity to link national REDD+ initiatives with local REDD+ projects. National REDD+ initiatives could be structured as enabling mechanisms that provide legal, technical, and financial support to local project activities. National REDD+ initiatives would shelter projects from national drivers of deforestation, nurturing the development of local projects and gradually populating the countrywide REDD+ strategy. Technical support to CF groups for carbon monitoring, certification, and verification could be included as key roles for national REDD+ managers. Such a relationship could dramatically reduce transaction costs, improve the quality of monitoring data, better coordinate mitigation activities, and accelerate the replication of projects.

In mid-2012, a re-measurement of 100 forest plots in Oddar Meanchey's REDD+ project was under way to assess the effectiveness of mitigation activities in controlling the drivers of deforestation. Updated satellite image analysis will also detect changes in forest cover and condition and the impact of communities in slowing deforestation in the project areas. Any achievements will be largely due to the voluntary efforts of forest-dependent communities, local monks, and NGOs who have actively engaged in forest patrols and awareness-raising over the past five years. Their effectiveness was due, in part, to support received from the FA that has encouraged the formation of CF groups and moved quickly to establish formal CF agreements. To that end, the Oddar Meanchey Project is demonstrating that an effective collaboration between forest-dependent communities and the FA can be supported through REDD+ project frameworks and the NCFP.

At the same time, the absence of financial support for field activities has constrained the rate at which this REDD+ project has developed, as have the inefficient methodological and certification procedures that characterize sub-national REDD+ design processes in many countries. Flexible funding sources for small community-based REDD+ initiatives and user-friendly methods, standards, and certification tools are urgently required to enable the development of more REDD+ projects in Cambodia.

The Oddar Meanchey Project has the potential to generate nearly $50 million over the next 30 years, assuming 7.1 million tCO2e (tons of carbon dioxide equivalent) at $7 per ton.[12] The extent to which it can achieve this level of REDD offset depends both on civil society and government partners as well as the viability of international carbon markets. The decision by the RGC to approve a REDD project in Oddar Meanchey as an alternative to leasing the same area for ELCs, was based on an intention to determine if REDD+ projects could conserve forests while generating capital flows and contributing to economic development. To succeed, the project will need to generate carbon sales which improve forest management and household livelihoods in the project area, while covering the costs of project operations incurred by the RGC and other stakeholders.

NOTES

1. "Proceedings of Meeting to Brief the International Community and Senior Officials of the Royal Government of Cambodia on Forest Policy Reform" (Phnom Penh: ARD, Inc./MAFF,1998).

2. Amanda Bradley, "Does Community Forestry Provide a Suitable Platform for REDD? A Case Study from Oddar Meanchey, Cambodia," in *Lessons about Land Tenure, Forest Governance, and REDD+*, ed. Lisa Naughton-Treves and Cathy Day (Madison, Wisconsin: The Land Tenure Center 2012), 68.

3. Ibid.

4. Letter from the Council of Ministers, Sar. Chor. Nor. No. 699, Council of Ministers, Royal Government of Cambodia, Phnom Penh, May 2008.

5. Joel Brinkley, "Cambodia's Curse," *Foreign Affairs* (March/April 2009), 113; Jefferson Fox, Dennis McMahon, Mark Poffenberger, and John Vogler, *Land for My Grandchildren* (Cambodia: CFI/EWC, 2008).

6. Personal communication, 2008.

7. NIS, "General Population Census of Cambodia 1998, Final Result" (Phnom Penh: National Institute of Statistics, Ministry of Planning, 1998), 299.

8. Top Neth et al., "Variation in Woodfuel Consumption Patterns in Response to Forest Availability in Kampong Thom Province," *Biomass and Bioenergy* 27 (2004): 61.

9. See Letter from the Council of Ministers, "Sar. Chor. Nor. No. 699."

10. Bradley, "Does Community Forestry Provide a Suitable Platform for REDD?," 67.

11. Ibid., 70.

12. Ibid., 69.

COMMUNITY FISHERIES IN TRANSITION IN THE TONLE SAP: A CASE STUDY OF KOMPONG PHLUK

MELISSA MARSCHKE

THIS CHAPTER WILL EXPLORE HOW COMMUNITY-BASED management projects have evolved, drawing on past and present experiences in one flood forest commune.[1] This helps to situate the successes and failures of community-based approaches within the Cambodian context, and illustrates how Kompong Phluk provides a strong case for the potential of community fisheries in flood forest areas. At the same time, I argue that fisheries committees and natural resource management generally may need to shift their role. Broader socioeconomic and environmental drivers create doubt about the long-term viability of small-scale fisheries and reinforce the need for alternative livelihood sources. That said community fisheries committees could play an important role in thinking through local responses to such traditions.

> I don't fish anymore, since I am now too old, but one of my sons (of 4) continues to fish. He likes fishing, and is doing all right because he also raises milk fish (a form of aquaculture). Some of my children have now moved to the city and some continue to stay here. In general, I think life is harder here...although it is quieter. My wife now mostly stays in the city, to sell goods from

one of my sons' home, whereas I sometimes help my son with fishing and am still involved in some community fisheries work too.

—Excerpt from an interview with Ros Norn, May 2010

The above comments from Ros Norn were made at the start of our discussion. It had been a few years since I had last visited Kompong Phluk, and my purpose in returning to the village was to see how people were doing and to investigate the status of the community fisheries and general resource management work.

However, just as the formal part of our interview began, the village chief came running down the main road, megaphone in hand, announcing that a fire was nearby. He told people to get their things to the river and to tear down all the thatch from their homes. We immediately climbed to the topmost point of Norn's home to see the location of the fire. It was not far away—plumes of smoke could easily be seen. The villagers immediately jumped into action, taking all of their valuables to the nearby river and dismantling their homes (both the thatch and the valuable wood). It was a frenetic few hours, with everyone helping each other out. People were understandably quite nervous, as the fire quickly moved to within two hundred meters of the homes, having burned through a rather large tract of flood forest. The fire trucks arrived just in time, to ensure all houses were hosed down with water and to keep the fire at a safe distance. It took several days for the fire to stop smoldering and for villagers to be able to return to reconstruct their homes.

I was not able to complete this interview, or any other interviews in the village. During the fire emergency, immediate welfare came first, so my research assistant Dyna and I helped out as best we could, mostly by ferrying things to the river with our motorbike. This experience really stuck with me. While villagers had talked of flood forest fires—in 2005, a forest fire burned down a large section of the village—this was the first time in all my years of fieldwork that I really felt the vulnerability that flood forest villages face in the hottest months of the dry season.

Fire is a constant threat, and its ability to destroy can be devastating, even more so because most people keep their money and other valuables in their homes, not in banks. This experience also helped me to understand why some people really do not want their children to become fishers in the Tonle Sap, as they see life in this area as being too risky. The extreme conditions that result in living in an area that is inundated with water half the year and the constant threat of forest fires in the dry season take their toll on most people.

Norn is from one of three villages found in Kompong Phluk, a flood forest commune (the three villages are interconnected, so I will refer to this area as Kompong Phluk) within the Tonle Sap basin that is known for its strong fishing roots. There is a history of flood forest and fisheries management here, dating back to the 1940s or possibly earlier, and this is an area that has been relatively successful at promoting natural resource management. In 2000, a community fisheries committee was established, and the area surrounding the commune (including forests inundated for part of the year) was recognized at the national level by the Fisheries Administration as belonging to local people for natural resource management. Several international organizations lent extensive support to villagers to help them to organize, plan for, and implement natural resource management and, in more recent years, ecotourism (tourists come to see the village and flood forest area).

Even with such support, it is not easy to protect the resources in this area, as demonstrated by the fire that I witnessed. In addition, if one puts in mind the declining fish stocks, the threat of encroachment by farmers wanting to turn flood forests into farmland, and the lure of economic opportunities elsewhere, one can understand why rural life in Kompong Phluk is changing. While this commune continues to thrive in many ways, particularly for those fishers who are able to combine fishing and aquaculture activities together, it is very much an area in transition. In this sense, Norn's household provides a good example of how some people are moving out of the fisheries occupation altogether, while others are finding ways to increase the viability of fishing through forms of small-producer aquaculture.

What, then, is the role for community fisheries in flood forest areas (see figure 1), particularly in the face of major socioeconomic and environmental change? Kompong Phluk provides an excellent case through which to explore this question, especially in terms of flood forest and fisheries management, since this commune is known throughout the Tonle Sap area for having protected much of their flood forests. At the same time, people in the commune are aware that life for their children may look different and may not necessarily involve fishing. Meanwhile, a series of reforms initiated at the national level over the past decade has led to a series of new policies that support local involvement in resource management and many donor projects that focus on this type of work. Community fisheries, community forestry, and others forms of "community" management all stem from a broader trend towards decentralized resource management. This form of local resource management is not adequate, however, for dealing with the emerging governance challenges found in the area, particularly in terms of regional hydroelectric development on the Mekong River, oil and gas exploration, and mining for sand.

Figure 1. Map of Tonle Sap Floodplain

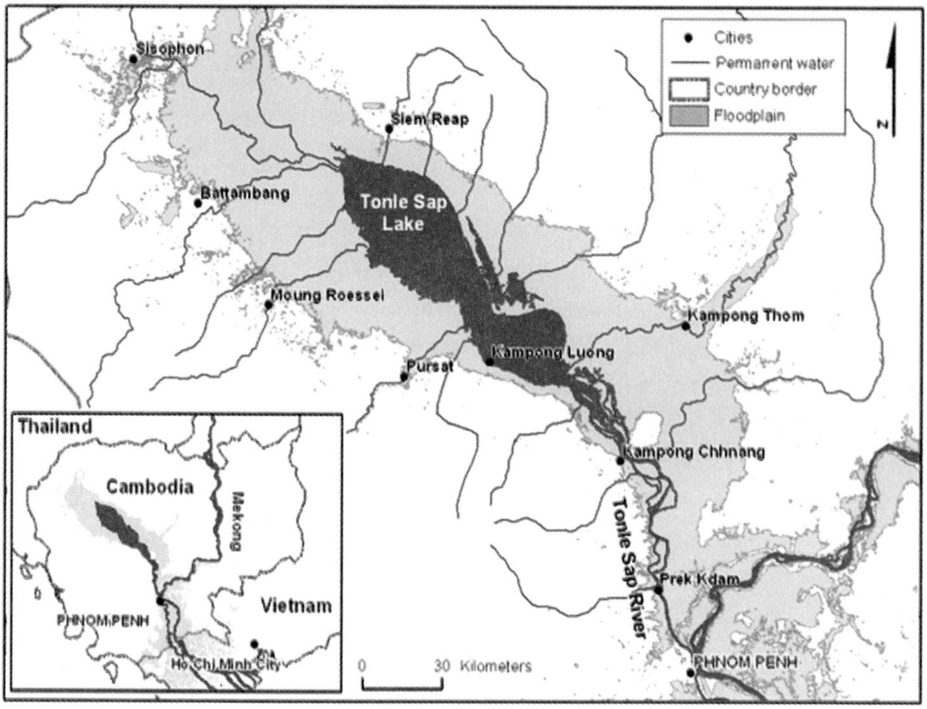

Source: Marschke 2005

Turning to the fisheries sector, which has jurisdiction over flood forests, there has been a growing recognition in the past ten to fifteen years that state-led approaches, represented in their simplest form as a combination of conservation objectives with policing tactics, have failed in both protecting fishery resources or in generating a shared interest in fisheries management among fishers and the state.[3] Meanwhile, significant fieldwork in the area of community forestry had been promoted since the mid-1990s and, by the late-1990s, there were early signs of success to extend community management to the Tonle Sap. The timing was right in terms of promoting local involvement in the management of a resource within another resource sector, namely fisheries, particularly since fishers were beginning to mention stock declines and conflicts over access to the fishery were increasing.[4] Thus, forms of local-government partnership (i.e., community-fisheries or fisheries co-management) became attractive to a number of actors.

Experiences like that of Kompong Phluk have been made possible by new laws and policies being implemented by the RGC. In 2000, Cambodia's prime minister initiated a reform of the fisheries sector when the central government scaled back the extent of individual fishing lots by releasing 56 percent of Cambodia's commercially zoned fishing area (around 5000 ha) for potential community fisheries.[5] This administrative reform led to significant changes, not only in terms of institutional reforms, such as the creation of a Community Fisheries Office or the promotion of community fisheries within Cambodia's Millennium Development Goals, but also changes at the village level that enabled local committees to form and work on local fisheries governance. A handful of community fisheries committees existed in the late 1990s, but by 2010 there were 468 community fisheries sites (433 inland, 35 coastal).[6] At this point, only 173 sites have been officially registered, even though the Community Fisheries Sub-Decree was passed in 2005 (to be recognized, communities need to work with fisheries staff to create maps, rules, and regulations for a specific area).[7]

According to the Community Fisheries Sub-Decree, community fisheries have five stated objectives, which include the following:

1) To manage inland fisheries and related ecosystems where fishing lots have been cancelled;

2) To manage fisheries resources in a sustainable and equitable manner;

3) To increase understanding and recognition of the benefits of ·
   fisheries resource through participation in management;

4) To provide a legal framework to establish community
   fisheries; and

5) To improve the standard of living and reduce poverty.

Villagers get involved in resource management (fisheries, forest protection, etc.) activities for various reasons. A newly elected resource management committee, for example, may enable actors to find a new patron
and serve as a way to alter village power relations.[8] This may also serve as a
mechanism to ensure that those who have been historically excluded from
natural resources near their villages (vis-à-vis fishing lots, forest concessions) can now gain access. A committee may also be organized as a way to
garner resources to stop encroachment from outsiders on local resources.
There may be strong leaders who are interested in protecting resources
near the village or villagers may feel that it makes sense to work on these
issues.

It may also be that the area has experienced significant declines, and
an opportunity through a technical department or through an NGO pilot
project enables the establishment of a community-based natural resource
management site.[9] Or, it may be that community members are interested
in remaining in rural areas and recognize that part of their livelihood is
linked to natural resources. Although Cambodian communities are no
longer particularly isolated, and household members are now moving in
and out of the village, rural livelihoods remain linked to ecosystems.

Snapshots of what community involvement in resource management
may entail do exist. A 2009 edited volume of case studies, entitled *Emerging
Trends, Challenges and Innovations: Community-based Natural Resource
Management (CBNRM) in Cambodia*, highlights cases of successful local
resource management in Cambodia. For example, while initial resource
management initiatives emphasized resource protection, this volume of
cases demonstrates how people can move beyond conservation to manage
and harvest their resources, such as honeybees and pole cultivation. Cases
also deconstruct ideas around common property, participation, and who
gains from an involvement in resource management. What is perhaps
less clear is whether these types of efforts can be realistically sustained

in a context such as Cambodia, particularly when donors pull back and the government remains largely responsible for supporting this type of effort. This is why reviewing Cambodia's past might provide hints about the viability of this effort in the future. The following provides a detailed examination of how fisheries and flood forest areas were managed in the Tonle Sap generally, and specific oral histories from elders in Kompong Phluk relating to management issues.

FISHERIES MANAGEMENT IN THE TONLE SAP

This section summarizes what is known about Cambodian fisheries governance since French colonization.[10]

- The Tonle Sap's aquatic resources have typically generated huge revenues for Cambodia's rulers and administrators, from the Cambodian royalty to the French colonizers to the present-day Fisheries Administration and high-level officials.

- The Tonle Sap Lake and its surrounding flood forest land were historically managed as a common resource for all. This began to change with the arrival of French colonizers, who divided up the Tonle Sap into privatized fishing concessions or "lots" auctioned off for commercial exploitation. This concession system continued through independence, until the rise of the Khmer Rouge (1975–1979).

- During the Khmer Rouge era, fisheries resources were neglected in favor of rice production; agricultural expansion in this era included the cutting of some flood forest areas.

- In the early 1980s, the Vietnamese-backed government introduced a type of collectivization whereby farming and fishing villages were organized into solidarity groups, known as *krom samaki*. However, these reforms were short-lived and, by the late 1980s, the government reverted to the old concession model, along with the introduction of liberal market reforms. Fishing lots in the Tonle Sap were auctioned off to the highest bidder (ranging from $2,000 to $20,000), thereby excluding villagers from traditional fishing grounds.

• Between 1980 and 1998, 50 percent of the flood forests around the Tonle Sap Lake were deforested, pointing to a trend of general resource decline found throughout Cambodia over the past twenty years.[11] Although there is mention of fish stocks declining throughout the years, this is also the first time that villagers really point towards serious declines. By the late 1990s, to the exclusion of many fishers, a small number of fishing lot operators controlled about 10,000 square kilometers of the most productive and lucrative fishing areas in the Tonle Sap. People were angry.

• Conflicts over the concession system multiplied between 1995 and 1999, in part because dissent was made possible in the context of expanded civil and political rights spelled out by the 1993 constitution. Escalating conflicts between local fishers and fishing lot operators brought fisheries management to the attention of international donors and Cambodia's politicians. The Fisheries Reform, initiated in 2000, emerged in response to these tensions and after pilot projects demonstrated the potential of local people to be involved in resource management.

LOCAL RESOURCE MANAGEMENT IN KOMPONG PHLUK

A few researchers, taking a historical perspective on natural resource management and local agency, found that local resource practices were likely led by village and higher-level government officials. In the coastal areas of Southwest Cambodia, as an example, there appears to have been a system in place for charcoal production, where charcoal was produced in designated places and a recovery period was recognized as necessary to ensure the longer-term sustainability of the mangrove ecosystem. This effort was supervised by local government officials.

In the 1960s, a state-controlled marketing system prevailed allowing villagers to do what they liked with their resources, as long as they sold their products to a centrally appointed government official. In this system, household harvests were recorded, and money was distributed after the district chief sold the harvest on behalf of households. To draw from an example from Kompong Phluk in the Tonle Sap, monks recall villagers protesting as parts of the flood forest near the village were cleared for watermelon cultivation in the 1940s. Such cutting affected aquatic habitat during the flood season. After complaints were lodged at the district level, watermelon cultivation stopped and reforestation was encouraged. Villagers then followed an informal system of forest management, reporting to authorities any illegal cutting or hunting activities; this informal system was eventually formalized.[12] The box below provides further details—its history of flood forest management in the area helps to at least partially explain its relative success today.

A HISTORY OF LOCAL MANAGEMENT IN ONE FLOOD FOREST COMMUNITY

1930s–1940s	Parts of flood forest near villages cleared for watermelon cultivation. Wind and storms began affecting these villages; cutting also affected aquatic habitat. Late 1940s, villagers complained to district level. Watermelon cultivation stopped, and reforestation was encouraged near villages.
1960s	Some forest cutting for mung bean cultivation; pumpkin, cucumber, and other vegetables are also grown near village. Dense forests found near villages.
Khmer Rouge	Mung bean cultivation expands to edge of the flood forest; other resources (fish and trees deeper in lake) not used.
1980s	Collective fishing (circle nets) and farming (mung bean farms divided among 300 village households); immediate forest surrounding village was "protected."
1987/1988	Administrative reshuffling: parts of Kompong Phluk commune transferred. District authority hands over mung bean farms to upland communes.
Early 1990s	Slash and burn of the flood forest area, especially bordering other communes, for conversion into mung bean farms. Significant conversion again affects aquatic habitat.
1994	Fire, possibly from "outside" fishers (upland areas), burned 200 ha of flood forest, enraging villagers and serving as a catalyst for further flood forest protection.

1998	The community requested assistance from NGO to expand the area of community "protected" forest and stop encroachment on mung bean farming; replanted seedlings to 50 ha of former farmland.
1999–	A committee elected, first to deal with flood forest protection, and then to expand community fisheries. Varied support from the Fisheries Administration; backstopping of issues from UN-funded project (until 2004). Relatively deep connections have been developed with government and NGO actors, which are drawn on in times of conflict or crisis.
2005–	Work begun on ecotourism (a specific ecotourism center now exists, with a small restaurant).
2010–	Committee continues to meet, perhaps less regularly, but does deal with pressing resource management issues.

Adapted and updated from: Evans et al. 2004; Marschke 2005

The box illustrates that villagers in Kompong Phluk were motivated to protect their homes from storms and winds; hence, flood forest protection practices emerged. Already in the 1940s, tension existed between farmers who wanted to clear flood forest land for agricultural production and fishers who depended upon the flood forest during the flooding season for fishing activities. One of the few benefits of the Khmer Rouge regime was that fisheries resources remained intact, and the flood forests surrounding Kompong Phluk were barely cut. In the 1980s, most people in Kompong Phluk farmed mung beans and relied on the fishery. With administrative reshuffling in the late 1980s, Kompong Phluk farmers could no longer access any farmland, and the residents became much more reliant upon the fishery.

Throughout the 1990s, upland farmers steadily encroached on Kompong Phluk's flood forest fishing grounds, cutting down flood forest trees as a way of staking claims to the area.[13] This was perceived to be affecting the aquatic habitat for a variety of fish species, since flood forest areas serve as excellent habitat, protectors, and breeding grounds. At the same time, villagers in Kompong Phluk focused on an informal system of flood forest protection, both as a means of protection from storms and as a way to ensure aquatic habitat and breeding grounds. During this time, community members used an informal monitoring system that enabled them to sometimes stop outsiders from cutting trees by issuing warnings and creating peer pressure to not cut trees.[14]

However, resource pressures continued to increase, which partly explains why villagers decided to take advantage of the opportunity to work with a UN-funded project in the mid-1990s. The purpose of this work was to enhance flood forest protection measures and address fisheries issues more seriously.[15] It was a mandate that villagers could pragmatically accept. Villagers have since created extensive management plans, complete with zoning, designating areas for firewood collection, creating fish breeding sanctuaries, posting signs, implementing an environmental education and fire prevention campaign, doing formal patrolling, and creating a small restaurant as a means of capturing profits from ecotourism.[16] Although the elected committee does not meet every month, it comes together as problems arise and links closely with the commune council. The Kompong Phluk Resource Management Committee has created an advisory role for a member from the commune council. This committee has also been active in advocacy campaigns for local rights at the provincial and national levels.

Kompong Phluk's history of fisheries management helps to explain why certain villagers were willing to engage more formally in resource management work—in the sense of working with NGOs and government officials—as such opportunities arose in the 1990s and 2000s. Kompong Phluk exemplifies a "success story" of community-based resource management, although this is not to suggest that the committee can handle or is interested in handling all issues (there has been some flood forest encroachment, for example, and the committee does not address issues of stolen fishing gear).[17] Nonetheless, Kompong Phluk's history of resource management suggests that there will be some continuity of resource management efforts, even if in different forms, since there has been a sustained interest in general habitat protection (i.e., flood forest) and ensuring access to the fishery.

Kompong Phluk is not the only commune that has created a community fishery committee on the Tonle Sap Lake. Several other villages or communes have also done so. In general, there is a perception that those communes or villages that have been more active in flood forest and fisheries habitat protection—through informal patrolling where villagers report illegal activities to the committee or through a committee organizing formal patrols—have managed to sustain more of their forests. There is no study to "prove" this, but Tonle Sap villagers and, more recently, NGO staff do talk about those areas where there is better protection. Additionally, one sees signposts throughout the flood forests, some with protection slogans and others warning that cutting is illegal. In some cases, this effort is linked to a strong local history of protection; in other cases, it is linked to strong leadership or an NGO that has worked steadily to promote the idea.

CHALLENGES AND CHOICES FOR THE GOVERNANCE OF THE TONLE SAP

Some of the challenges facing organizations and fisheries committees working in the Tonle Sap are broader resource governance challenges. Even if specific fisheries or flood forest conflicts were to be solved, there are other issues that require higher levels of governance (national and regional) to solve. Table 1 highlights a few of the threats facing those living in the Tonle Sap region.

Table 1. Threats to the Tonle Sap

Threats	Explanation
Climate change	Rise in temperature may affect agricultural activities, thereby placing more pressure on the fisheries. A rise in temperature may affect migration routes of certain fish stocks. Acidification may also be an issue.
Damming the Mekong River	Approved hydroelectric dams in the upstream Mekong area will greatly impact the entire Tonle Sap basin, dramatically altering rural life.
Oil and gas exploration	In 2009, a state-owned Vietnamese company signed a 7-year contract with Cambodia's National Petroleum Authority to explore oil and gas in the Tonle Sap. Besides exploration disruptions (fish stocks, etc.), it is unclear what this development might mean for the flood forest fishing villages.
Deforestation	This is linked to economic development generally (roads), and to agricultural expansion specifically (mung beans, etc.). As areas become further enclosed, the flood forest becomes more and more appealing to multiple actors.

Aquaculture development	There are pros and cons here, as this may become an important income source. As areas become enclosed for aquaculture production, what this means for fishers who cannot enter aquaculture is less clear.
Overexploitation of fish and wildlife	If fishing is declining as suggested, it appears to be a "race to the last fish" scenario unless management is strengthened throughout the entire Tonle Sap, and larger-scale fishing is monitored. Wildlife is also being exploited.
River mining	Although mining in rivers was banned in 2010, the practice continues. This alters fish habitat and can lead to significant habitat destruction.

While villagers in the Tonle Sap are well-adapted to seasonal change, and have fostered a series of adaptation strategies that include seasonal migration in the Tonle Sap and innovative housing designs for dealing with seasonal floods, people are vulnerable to ongoing stresses (climate-induced) and shocks (damming on the Mekong). This vulnerability is further impacted by poverty levels in the Tonle Sap and consistent stress that is experienced by the fishers, such as fires and overfishing.[18] Villagers themselves report changes to flooding seasons, intensive drought, decreasing water quality, and declining availability of fish and other natural resources.[19] The scale of challenges now facing the Tonle Sap is grave. Besides climate-induced shifts, there is a general trend of declining resources, competition from aquaculture, and economic development projects such as sand mining, dam construction, and oil and gas exploration.

Added to these "push" factors are "pull" factors, such as the potential of employment in the city or in neighboring countries. Since most villagers do not have English skills, they are not able to access tourist-related jobs, but work in photocopy shops as motortaxi drivers or as cleaners. A few villagers can access garment factory jobs, although one generally needs a "connection" to do so. Construction work or serving as a laborer on fishing boats in Thailand are big draws, although reports of labor conditions can be horrific.[20] Additionally, these villages are often isolated from provincial towns, relying solely on boat transportation and seasonal road access in some cases.

Although a few wealthy households choose to invest in their children's education, this is not an option available for many households. The future of capture fisheries is called into question, as resources decline and young people look to alternative sources of livelihoods. In some cases, households transition into aquaculture, whereas in other cases, people leave the area. Tellingly, community leaders emphasize the need for young people and poor households to be able to access alternative livelihood sources and acknowledge that rural life is shifting.

Government actors also acknowledge that the Tonle Sap region is an area in transition. While the Tonle Sap has been seen as Cambodia's food basket and as a major revenue source for centuries, this may be shifting depending on how small-producer aquaculture is managed. The Fisheries Administration, for example, acknowledges the various ways in which natural resources and the current environment might be altered, including through aquaculture, drawing on the Mekong as a source of hydropower, and turning flood forest areas into roads or reservoirs.[21] At the same time, the NSDP (National Strategic Development Plan) acknowledges that poorer households depend on a diverse natural resource base to survive in this region, and that health, education, and good governance are all keys to the Fisheries Administration's longer-term strategy.

Acknowledging this would seem to suggest a full understanding of shifts in rural life and competing interests in natural resources. Government and NGO-led approaches need to see the various economic opportunities found in the Tonle Sap, consider which projects pose the greatest harm to the ecosystem and rural poor, and make careful choices. Fishers on the selling end of the value chain do not have the advantage in economic relations, and this needs to be addressed.[22]

The trade-off between economic development (e.g., dam construction for electricity generation) and capture fisheries merits further exploration.[23] Advocates of dam construction argue that harnessing the Mekong River's great potential for hydropower generation will contribute to poverty reduction and sustainable economic growth.[24] Yet, dam construction in China, Vietnam, and Laos has negatively impacted communities living downstream along the Mekong River, including those in Cambodian villages. Adverse impacts include changes in the river's hydrology and water quality that threaten to increase poverty, health problems, and food insecurity for affected communities.[25] For instance, construction of the Yali Falls dam, located on Vietnam's Sesan River, approximately eighty kilometers from the border of Cambodia, resulted in massive flooding, the drowning of thirty-five people, and the destruction of homes, livestock, and property in Cambodia. This experience does not bode well for Tonle Sap villages.

CONCLUSION

Community fishery management committees were formally created in the 1990s, resulting from tensions over access to fishing grounds, the state's interest in reforming the fisheries sector and involving local people in such a reform, and support from international donors in funding decentralized resource governance initiatives. Committees were elected to protect and manage aquatic and forest resources. Active committees have ensured some flood forests remain in and around their villages—this is no small feat. What is less clear is how long these committees can sustain such protection of flood forests or local fishing grounds, given the development pressures found in this region.

This is why rethinking the role of local committees is necessary. Such groups may be called to play a wider role to ensure the future of capture fisheries and rural development in the Tonle Sap. Declining fish catches, climate change, conflicts between user groups, and competition from economic development projects are real challenges that cannot be addressed at the local level alone. However, since these committees represent their respective villagers and have been working on resource management issues over the years, they are well-placed to advocate with higher authorities for their fishing villages and to contribute ideas on how to best incorporate natural resource management in this ever-shifting landscape. Committees are also in a position to help villagers, especially resource-poor ones, gain access to alternative livelihood sources, and to advocate for adaptation strategies that can enhance rural life in the face of multiple environmental changes.

Diversification into activities outside of fishing (small-producer aquaculture, home business, sending a family member to the city), as the quote from Norn at the beginning of this chapter highlights, is already happening. This also suggests that the role of capture fisheries needs to be rethought. Capture fisheries might not be considered as a short- to medium-term strategy. For now, capture fisheries should continue to be a central element of poverty alleviation strategies, along with forms of local resource management. However, these discussions should be tempered by knowledge about the important obstacles facing the fisheries sector and about rural shifts in general. For example, casting the role of flood forest areas in terms of avoided deforestation and linking into payment strategies for environmental services require careful consideration.

In the long term, it may be that flood forest protection has the greatest chance for success if incorporated into payment for ecosystem services regimes. Fisheries management, on the other hand, cannot only be left to villages; rather, the Fisheries Administration and the donor community need to ensure general policy uptake and enforcement. Finally, the potential for small-producer aquaculture to be a rural livelihood activity requires far greater thought. Given the importance of fish as a source of dietary protein and the growth of aquaculture in the region, this could play an important role in poverty alleviation, if managed properly.[26] Local fisheries committees, such as Kompong Phluk's, could play an important role in guiding and managing this transition process.

NOTES

1. A few sections found in this chapter are based on my forthcoming book *Resource Governance at the Margins: Fish, Trees and Life in Coastal Cambodia* (University of Ottawa Press). Fieldwork in the Tonle Sap was conducted between 2002 and 2004 (see Marschke 2005), with follow-up visits to Kompong Phluk in 2007 and again in 2010. Many thanks to Dyna Eam and Kim Nong who have consistently accompanied me to Kompong Phluk over the years. Three of my undergraduate students have helped tremendously in pulling together this chapter. Thanks to Mia Choinière and Sarah King for gathering background materials on the Tonle Sap and for creating several of the tables found in this chapter. Thanks to Emily Rees for conducting a handful of interviews with those working on community fisheries issues in Cambodia in March of 2011.

2. Kompong Phluk is a flood forest area that is innundated for part of the year. This phenomena happens in monsoon countries, such as Cambodia; the Tonle Sap is the largest freshwater lake in Southeast Asia. During the wet season (June to October), the Mekong River rises and the water inundates the dry land, flooding the forested areas. As the river rises, many fish migrate to spawn in the protected, nutrient-rich waters surrounding the flood forests. This is how the Tonle Sap expands four to five times its dry season size, increasing in average depth from one meter to 7–9 meters. When the water recedes (November to May), it drains, revealing the forest once again and stranding fish in pools and ponds (Evans et al. 2004; Marschke 2005). This area holds a rich biosphere of more than 200 species of fish, 42 types of reptiles, 225 species of birds, and 46 kinds of mammals. Cambodia's flood forests provide firewood and building material for villagers and are vital breeding grounds for fish. At the same time, flood forests are threatened by deforestation, conflicts, poorly enforced policies, etc. Villages found in the Tonle Sap area, therefore, are particularly adept at adapting to shifting seasons and to thinking about resource management in a rather holistic manner since flood forest protection is an essential part of any community fisheries strategy in these areas.

3. N. Nasuchon and A. Charles, "Community Involvement in Fisheries Management: Experiences in the Gulf of Thailand Countries," *Marine Policy* 34 (2010): 163–69; R. Arthur, R. Friend, and M. Marschke, "Making Adaptive Co-management more than a Marriage of Convenience: Reconciling Theory and Practice in the Management of Fisheries in the Mekong Region," in *Collaborative Resilience: Moving From Crisis to Opportunity*, ed. B. Goldstein (Massachusetts: MIT Press, 2011).

4. P. Evans, M. Marschke and K. Paudyal, *Flood Forests, Fish and Fishing Villages, Tonle Sap, Cambodia: Community Forest Management Trends in Southeast Asia* (Manila: Asia Forest Network, 2004).

5. C. Sneddon, "Nature's Materiality and the Circuitous Paths of Accumulation: Dispossession of Freshwater Fisheries in Cambodia," *Antipode* 39, no. 1 (2007): 167–93.

6. *National Strategy Development Plan, 2009–2013* (Phnom Penh, Cambodia: Fisheries Administration, 2010).

7. *Emerging Trends, Challenges and Innovations for CBNRM in Cambodia*, CBNRM 2nd ed. (Phnom Penh, Cambodia: CBNRM Learning Institute, 2009).

8. J. Legerwood and J. Vijghen, "Decision-making in Rural Khmer villages" in *Cambodia Emerges from the Past: Eight Essays*, ed. J. Legerwood (DeKalb, IL: Center for Southeast Asian Studies, Northern Illinois University, 2002), 176–223.

9. M. Marschke and J. Sinclair, "Learning for Sustainability through Participatory Resource Management," *Journal of Environmental Management* 90, no. 1 (2009): 206–16.

10. P. Degen, F. van Acker, N. van Zalinge, N. Thouk, and D. Loeung, "Taken for Granted: Conflicts over Cambodia's Freshwater Fish Resources" (Paper presented at IASCP Bloomington, Indiana, June 2000); M. Slocomb, "The Nature and Role of Ideology in the Modern Cambodian State," *Journal of Southeast Asian Studies* 37, no. 3 (2006): 375–95; J. Kurien, N. So, and S.O. Mao, "Cambodia's Aquarian Reforms: The Emerging Challenges for Policy and Research," in *Land Fisheries Research and Development Institute* (Phnom Penh, Cambodia, 2006); S. R. Bush, "Contextualising Fisheries Policy in the Lower Mekong Basin," *Journal of Southeast Asian Studies* 39, no. 3 (2008): 329–53; M. Marschke, *Resource Governance at the Margins: Fish, Trees and Life in Coastal Cambodia* (Ottawa, Canada: University of Ottawa Press, forthcoming).

11. Evans, Marschke and Paudyal, "Flood Forests."

12. M. Marschke, Le Secteur des Pêcheries de l'Ère « Post » au Cambodge. Une explication de la non-transformation, *Anthropologie et Sociétés* 32, nos. 1–2 (2008): 133–54.

13. M. Marschke, "Livelihood in Context: Learning with Cambodian Fishers" (Ph.D. Thesis, University of Manitoba, 2005).

14. Evans, Marschke and Paudyal, "Flood Forests."

15. M. Marschke and F. Berkes, "Local Level Sustainability Planning for Livelihoods: A Cambodian Experience," *The International Journal of Sustainable Development and World Ecology* 12, no. 1 (2005): 21–33.

16. M. Marschke and F. Berkes, "Exploring Strategies that Build Livelihood Resilience: A Case from Cambodia," *Ecology and Society* 11, no. 1 (2006): 42. [online] http:www. ecologyandsociety.org/vol11/iss1/art42.

17. M. Marschke, Le Secteur des Pêcheries de l'Ère Post au Cambodge, 133–54.

18. World Bank, "Cambodia: Halving Poverty by 2015" (Phnom Penh: World Bank, February 2006).

19. P. Nuorteva, M. Keskinen and O. Varis. "Water, Livelihoods and Climate Change Adaptation in the Tonle Sap Lake area, Cambodia: Learning from the Past to Understand the Future," *Journal of Water and Climate Change* 01, no. 1 (2010): 87–101.

20. M. Marschke, interviews, 2010.

21. *National Strategy Development Plan, 2009–2013* (Phnom Penh, Cambodia: Fisheries Administration, 2010).

22. V. Loc, S. Bush, L. Singh and N. Khiem, "High and Low Value Fish Chains in the Mekong Delta: Challenges for Livelihoods and Governance," *Environment, Development, and Sustainability* 12, no. 6 (2009): 889–908.

23. R. Arthur, R. Friend, and M. Marschke, "Making Adaptive Co-management more than a Marriage of Convenience: Reconciling Theory and Practice in the Management of Fisheries in the Mekong Region," in *Collaborative Resilience: Moving From Crisis to Opportunity,* ed. B. Goldstein (Massachusetts: MIT Press, 2011).

24. J. Bird, "A Responsible Approach to Building Dams on the Mekong," *The Nation,* September 22, 2008. Retrieved online 2011:
    http://chrislang.files.wordpress.com/2008/11/the_nation_22_sept_08.pdf.

25. A. Trandem, "A Vietnamese/Cambodian Transboundary Dialogue: Impacts of Dams on the Se San River," *Development* 51 (2008): 108–13.

26. *The State of the World Fisheries and Aquaculture 2010* (Rome: Food and Agricultural Organisation of the United Nations, 2010).

LAND USE AND FOREST CHANGE IN RATANAKIRI

JEFFERSON M. FOX
MARK POFFENBERGER

LIKE MANY COUNTRIES IN SOUTHEAST ASIA, CAMBODIA FACES challenges respecting the rights and culture of its upland dwelling ethnic minorities while pursing national development strategies.[1] Centrally designed planning and economic goals have been prescribed for these remote areas, but often without recognizing the extraordinary knowledge indigenous communities have of their environment and the special resources they can bring to its further development. As a consequence, public and private sector initiatives for development may fit poorly, or conflict with local needs and management systems, resulting in destabilizing shifts in land use and tenure systems, as well as social systems.

Ratanakiri Province is located in a remote northeast corner of Cambodia. The province's 250 villages are heavily dependent on forests, with 100,000 people who live either within the forests or within five kilometers of them.[2] Annual population growth of 4 to 5 percent from natural increase and migration, combined with rapidly expanding market penetration, is putting immense pressure on land and forests and fueling a large and illegal land market. As indigenous communities lose control of their lands, they are forced to retreat further into the forest, clearing those areas in turn. At the current rate of forest loss, it appears that much of the forest in Ratanakiri will be cleared in the next decade. During the same period, it is likely that half of all indigenous lands in the province will be transferred to outside investors, concessionaires, or Khmer migrants from lowland areas. The alienation of indigenous community lands is leading to growing social and economic marginalization, while the clearing of natural forests will likely destabilize microclimatic patterns, affect watershed hydrology, and erode biodiversity. These changes, in turn, may undermine the sustainability of new economic production systems that replace existing land use patterns (i.e., forests and swiddens).

This chapter draws on case studies from three communities in Ratanakiri to illustrate both the forces driving land use and tenure change, as well as how effective community stewardship can guide agricultural transitions. The researchers analyzed how indigenous communities that had historically managed forestlands as communal resources are responding to market forces and pressures from land speculators. Krala Village received support from local NGOs to strengthen their community, map their land, demarcate boundaries, strengthen resource use regulations, and develop land use plans. The two other villages, Leu Khun and Tuy, each received successively less support from outside organizations for purposes of resource mapping, and virtually no support for institutional strengthening. The remote sensing data indicates that in Krala, over the sixteen-year study period, protected forest areas remained virtually intact, while total forest cover declined at a rate of only 0.86 percent per year. By contrast, in Tuy Village, forest cover declined at a rate of 5 percent a year due to weak community leadership and institutions and heavy pressures from market forces (see table 1).

While under mounting pressure, the study finds that some indigenous resource management systems operating in Ratanakiri, like those in Krala Village, have demonstrated a capacity to achieve national goals for sustainable use and forest conservation. These systems respond well to support that is directed towards building local forest management initiatives and assisting traditional communal tenure. The study also indicates that indigenous families are under tremendous pressure to illegally sell community forests and are often manipulated by local officials. Indigenous community forestry presents an opportunity for the Royal Government of Cambodia (RGC) to retain high-value natural forests in Ratanakiri, if government, NGOs, and donors can find ways to effectively support traditional forest stewardship systems. Such a strategy would support the RGC's national forest cover goals, while responding to the social needs of the province's predominantly rural population.

EXPERIENCES WITH LAND USE AND TENURE CHANGE: THREE COMMUNITY PROFILES

Three villages were selected to present a continuum from least to most change undergone in the past ten years (table 1). These communities have all experienced rapid population growth and increased exposure to market forces over the past twenty years, yet their abilities to retain control over their resources, both in terms of tenure rights and land use, are dramatically different.

Table 1. Land Use and Tenure Change in Three Ratanakiri Communities

Event	Tuy Village, Ting Chac Commune, Bar Kaev District	Leu Khun Village, Ke Chong Commune, Bar Kaev District	Krala Village, Poey Commune, O Chum District
Extent of land tenure change	Moderate to High	Accelerating	Little to None
Rate of land use change	High	Moderate	Moderate
Village resettled (after Khmer Rouge)	1982	1979	1984
Population, 1979–1984	210 people	250 people	235 people
Population, 2007	458 people	639 people	420 people
Ethnic group	Tampouen	Jarai	Kreung
Estimated % of village land acquired or taken by outsiders	50%	10-20%	0%
Rate of deforestation, 1989–2006	5.0% per year	1.63% per year	0.86% per year

Tuy Village

Tuy is a predominantly Tampouen village located along the main road (Road 78) between the provincial capital of Banlung and the Vietnamese border, approximately twenty kilometers east of Banlung. The research team selected Tuy to represent "high land use and tenure change," based on reports that extensive land sales were taking place in the village and the surrounding communities. While the village has existed in its present location for many generations, the entire community was forced off their land in order to work lowland rice areas during the Khmer Rouge period. In 1982, approximately 85 families returned to resettle the village. By 2005, the population had grown to 103 families (458 people), at which time 23 of the families chose to break away in order to form a new village (Trang Village) as a result of internal conflicts between two community leaders.

Tuy's forests were logged extensively from 1985 to 1989 by Vietnamese companies, and later by the Cambodian military from 1990 to 1993. Since 2000, Tuy has seen significant changes not only in land use, but also in land tenure, with estimations that more than half of the community's productive land has been acquired by outsiders. People in Tuy increasingly see land and forests as market commodities and indigenous institutions as having diminished ability to guide community policies and behavior.

Leu Khun Village

Leu Khun is a Jarai village that was reestablished in 1979, when community members returned from lowland areas where the Khmer Rouge had forced them to relocate. At that time, there were seventy families who resettled the village, with a total population of 250. The population has since grown gradually to 130 families and 639 people. From 1986 to 1992, Vietnamese logging companies felled much of the larger, old-growth forests surrounding Leu Khun. Much of the remaining forest was felled by the Cambodian military throughout the 1990s, ending around 2002. Smaller-scale illegal felling continues. Leu Khun represents an established indigenous community that is coming under growing pressure from land speculators and where land use change is accelerating. Community members are uncertain how to address these problems, as village leadership is unable to deal with land speculators and corrupt officials, and some villagers see opportunities to generate cash through land sales.

Krala Village

Krala is a Kreung village of about 420 people in O Chum District, located about twenty-five kilometers north of Banlung. Krala was reestablished immediately following the fall of the Khmer Rouge, and the current settlement area of the village was established at its present site in 1984. During the Khmer Rouge regime, much of the community's population was relocated to Veun Sai District, where they were forced to farm paddy rice.

Unlike many of the indigenous communities adjacent to major roads in the province, Krala has managed to maintain control over 100 percent of its traditional land, and stands as a model for other communities who are facing a similar struggle. While the ownership of Krala's traditional land has not changed, the use of their land has seen a significant transition from entirely swidden agriculture in the early 1990s to the current mosaic of swidden fields and cashew plantations. In 1994, only four families in the village had planted cashew trees, but by 2000, nearly 100 of the 135 families in the village had planted cashews on their land. Now, it is estimated that every family in the village has at least 0.5 hectares of cashew trees. Krala has been the focus of several prior research studies that have resulted in a considerable amount of NGO attention and support in the village. It was selected to represent the community with the least amount of change (at least with regard to tenure), as well as a community where indigenous community institutions remain in control of communal lands, with support from local NGOs.

These communities provide insights into the experiences that indigenous villages are having as they confront the outside world and its market-driven economy. Krala, with the least outside pressure and most support at strategic times, has been able to retain its traditional culture and institutions more effectively, using its strengths to stabilize the community, while taking advantage of new cash crop opportunities. By Contrast, Leu Khun, under greater pressure and without adequate support, is reacting to the same changes, but its indigenous leaders and institutions are having greater difficulties formulating community-based natural resource management policies to guide land use and tenure transitions. Tuy has simply been swept away by more powerful political and economic forces, and it is foreseeable that it will gradually lose its identity as a Tampouen community. The next section describes some of the changes in land use and tenure that these villages have experienced from 1989 to 2006, based on remote sensing data and in-depth interviews.

TRADITIONAL TENURE AND CHANGE

The ethnic communities of Ratanakiri organize themselves in self-governing villages. Their strong social cohesion is built both on kinship ties, as well as through each family's sense of membership in the village, often linked to a belief in the spirits of their village. Among all groups, family elders are leaders held in great respect. They also play an important role in orchestrating land use decision-making. The elders' knowledge of customary law is of special importance for land and resource use, including their ability to deal with the power of the spirits of the forest, lands, and waters. Customary law, informed by the elders, governs and guides community decisions regarding the clearing of forests for agriculture.

According to customary law, the family has use rights over the land that they currently cultivate and over produce from old plots that are fallow, but may be farmed at a later date.[3] It is particularly important to note that customary law dictates that land use and control are generally passed along the female lineage. Under this system, the male will generally move in with the family of the bride, and will ultimately rely on the land and resources of that family for future cultivation. The close association with the land and the natural environment is reflected in the spiritual beliefs of each group, wherein the entire world is populated by spiritual forces associated with distinct components of the natural environment, such as trees, hills, stones, and water.

Substantial changes have taken place in land tenure between 1980 and 2007. Historically, the lands of Ratanakiri have been occupied and utilized almost exclusively by seven indigenous tribal groups. For the most part, these groups have commonly accepted territories where they practice long rotation farming, forest gathering, hunting, and maintaining their settlements. Land and forests were typically held as communal property with no practice of selling or transferring land. Instead, land was viewed as a resource held in trust by the community for future generations. The limits of each village's domain were traditionally understood and accepted on the basis of the outer extent of land cultivated by each community's members, whereby it is considered inappropriate to clear and cultivate land that must be accessed by regularly crossing land under cultivation by a neighboring village. In the past, the abundance of land options precluded any major disputes between neighboring villages.

In recent decades, the national government has begun to exert its claims to indigenous lands as part of the state's public land domain. After independence from France in 1953, the Royal Government of Cambodia (RGC) attempted to extend road infrastructure into the region, to build some educational facilities as a means to "Khmerise" the population, and to increase health services.[4] Aside from government efforts to integrate the province through projects and investments, spontaneous migrations from other parts of Cambodia are rapidly changing the social and ethnic composition of Ratanakiri.

In Ratanakiri, the population has expanded rapidly from 94,243 people in 1998 to 124,403 in 2005, with an expected population of 181,864 by 2013,[5] nearly doubling in fifteen years. An increasing proportion of the growing population is composed of migrants; consequently, the percentage of indigenous people fell from 68 percent in 1998 to 57 percent in 2005. By 2013, indigenous people will likely be a minority within Ratanakiri, their ancestral homeland. The combination of the growing number of migrants and outside investors is intensifying land competition in many parts of the province, while land is increasingly being viewed as a market commodity, even by local people.

While government demarcation of most land has yet to take place, technical agencies and planners have allocated economic concessions in the area for over a decade, including those for logging, mining, and other developments. Further, much of northern Ratanakiri was declared to be part of Virachey National Park (VNP), a 338,000-hectare protected area under the authority of the Ministry of Environment. In VNP, the creation of the protected area has resulted in the resettlement of indigenous Brao communities to areas outside the park.[6] In other parts of the province, there has been some displacement of local populations due to logging concession and mining activities. Nonetheless, with the exception of the Khmer Rouge Period (1970–1979), national land policies and projects have not yet been a cause of major indigenous population resettlement in Ratanakiri. However, these policies have created a growing awareness among local people that their land rights are being contested by the government. Indigenous land alienation is likely to accelerate in the future as national development plans are implemented in the region.

In addition to land claims made by the national government, land speculators and investors have moved quickly into the province to secure land and forests. Sales of indigenous land held under communal management are "illegal land transactions" under the national Land Law, which prohibits the sale of indigenous land. In relation to the land rights of indigenous communities, Article 25 of the Land Law states:

> Article 25: The lands of indigenous communities are those lands where the said communities have established their residences and where they carry out traditional agriculture. The lands of indigenous communities include not only lands actually cultivated, but also includes reserves necessary for the shifting of cultivation which is required by the agricultural methods they currently practice and which are recognized by the administrative authorities.

As discussed in sections below, some areas of mature forest may be included in the communal land title of indigenous communities. The possibilities for indigenous communities include communal ownership as described in Article 26 of the Land Law:

> Article 26: Ownership of the immovable properties...is granted by the state to the indigenous communities as collective ownership. This collective ownership includes all of the rights and protections of ownership as are enjoyed by private owners. But the community does not have the right to dispose of any collective ownership that is state public property to any person or group.

While community lands cannot be legally sold, factors such as corruption and the lack of surveys, registration, and documentation make indigenous lands vulnerable to speculators, who frequently enlist local officials to facilitate illegal sales. This is creating a rapidly expanding illegal land market, with indigenous communities increasingly aware that their communal resources have become a market commodity and a source of cash. The need for cash to meet education and health costs, improve housing conditions, buy consumer durables, and meet rising community and family expectations is common throughout virtually all indigenous communities in northeast Cambodia, so the motivation to raise cash through land sales is clearly present. This is abetted by an atmosphere of self-interest that culminates in minimal adherence to laws.

In most cases, community members are uncertain as to who owns title to the individual parcels that have been sold, as many of the initial transactions are made with a broker or land speculator who, in turn, sells it to other Khmers. Once the initial transaction is made, community members are no longer included in the process and may only learn about the final owner when development begins on the purchased land. The lack of transparency and clear communication in this process of multiple land transfers is creating tensions within villages. However, these offenses usually go uncontested, owing largely to the lack of process, documentation, and viable enforcement. As a result, feelings of discontent and animosity linger on, and some community members have suggested that there could be violence against these illegal developers in the future.

When asked about how land sales are recorded and parcels delineated, villagers are unable to define the process or any clear distinction on the limits of the land sold. In many cases, a commune official stamps a document noting the sale. Money is exchanged with a general understanding of transfer of ownership, without any surveying or physical demarcation of the limits. In some cases, the contracts are blank and verbal assurances are given that the document will be filled in later, according to what has been agreed upon. The villager may use his thumb print to notify approval of the sale, but rarely if ever receives a receipt or copy of the sales document. Moreover, because there is some shame associated with the practice of selling land, these transactions generally do not involve witnesses.

A devastating result of this lack of transparency and documentation is the common scenario where the new owners of the land will clear and use significantly more land than was originally agreed upon. When villagers protest or confront the new landowner about their breach of agreement, the landowner will frequently ask, "Where is your documentation to prove how much I purchased? Where is your stamp?" In the face of these responses, community members express a sense of confusion and powerlessness, feeling that they have no recourse for contesting their claim.

Once the land has been cleared and planted on by new owners, community members do not feel that they have the power or support to reclaim that land.

TRADITIONAL LAND USE SYSTEMS AND CHANGE

The entire highland way of life is inextricably linked to the forests and lands that surround, protect, house, and feed the communities. Perhaps most notable with regard to their relation to the forest and the land is the swidden agriculture technique and the practice of upland rice cultivation, which is common to all highland cultures. In this system, individual farmers clear a patch of regrowth forest for farming, where they will grow largely upland rice and some secondary crops like tubers, corn, and vegetables. They will grow upland rice (which can be planted and cultivated without tillage or major soil disturbance) from two to five years, at which time they will leave the land for fallow and begin clearing and farming a new patch of regrowth forest. During the fallow period, they will continue to harvest fruit and other wild vegetables that the old fields produce. While the swidden system of agriculture has been criticized as being inefficient and environmentally damaging, many researchers have shown that this practice has been highly sustainable, owing to low population densities and long fallow periods that allow substantial regeneration of the forest.

The hill tribe communities have sophisticated knowledge of local ecosystems and traditionally manage highly complex farming systems. The Brao-Kavet people in Kok Lak District are reported to cultivate 181 different crops in their swiddens, including 36 varieties of rice and 145 perennial and annual crops. A family swidden plot may include three to seven varieties of rice and sixty to one hundred other food species. Brao-Kavet people also recognize over one hundred "habitat" types, including wetlands, agricultural lands, and a wide range of forest types. Forests that compose a swidden system represent a range of age classes based on length of fallow. Forest classifications in the Brao-Kavet system include six stages of natural regeneration, from new fallows to old forests with very large trees.[7] While swidden farming is still practiced in many places in Ratanakiri, agricultural systems are in transition.

Land use practices are changing rapidly in most indigenous communities in Ratanakiri. Part of this reflects a broader agricultural transition that has been occurring in the uplands of Southeast Asia for decades. Traditional forms of subsistence agriculture that relied on a cycle of farming followed by lengthy fallow periods are being replaced by sedentary, market-oriented farming systems. While a few rubber estates were established in Ratanakiri during the colonial period, cash crop farming by indigenous communities has largely emerged since 1993, when Cambodia opened up for international investments and new road networks began to reach further into rural Ratanakiri.

Driving forces that affect all three villages include national policies to liberalize trade and markets and high market prices for rubber and cashews. Annual population growth in all three villages over the last fifteen years was relatively the same, ranging from 3.9 percent in Krala to 4.69 percent in Tuy (table 1). This suggests that population pressure may have been one of the forces driving land use intensification in Leu Khun and Tuy, and perhaps to a lesser extent in Krala. Other forces that affected the villages differently include the development of the road infrastructure that made it much easier to get to Tuy than to the other villages, illegal logging that occurred primarily in Tuy and Leu Khun, and the active engagement of the Non-Timber Forest Products Organization and other NGOs in assisting villagers in Krala to develop land use plans and to promote education.

Signs of an emerging cash economy are abundant within the three villages. Many families possess motorbikes, and several have televisions and VCD players, which they power with batteries that are charged in Banlung. Villagers also collect money for sacrifices and other community needs. In Leu Khun, one villager explained that when the water pump breaks, leaders collected 1,000R from each family to get the pump fixed. She said that people generally do not complain or resist making such contributions.

Finally, the need for cash has grown rapidly in indigenous villages in response to new opportunities to educate children. Some parents noted that an education is necessary in order to be able to negotiate better prices for goods in the market, and to be able to talk with government officials. Yet, the level of education available within the study communities was low. Many students attend schools in surrounding towns, but those without relatives to stay with face significantly higher costs.

In all three villages, a major driver of land use change is the growing reliance on cashew nut production as the primary source of cash income. While members of all three communities continue to rely on upland rice farming as their primary means of food production, virtually every family depends heavily on profits from cashew sales to supplement their family's food needs. Overall, people from all three villages indicate that they are in a better economic position today than they were in the 1990s, and they point to cashew production as the primary factor for this relative increase in prosperity.

The common scenario within all three communities involves an integration of upland rice cultivation with the establishment of cashew plantations, whereby rice is intercropped with cashew trees for a period of three years, when the trees mature and become productive. As the trees reach maturity and preclude the continued cultivation of rice, the common practice is to then clear an additional field, or extend the current field, and begin the process again.

While this basic scenario of land use change is equally descriptive of all three study communities, the most striking distinction among them can be seen in the level of coordination, planning, and overall awareness of the need for setting limits and maintaining portions of the land for swidden agriculture and forest conservation. Krala—having been the focus of less external pressure, stronger traditional leadership, and more intensive NGO support—has evolved a strong management structure, along with a clearly defined approach to land use planning. As such, each member of the community is highly aware of his or her rights to land, as well as responsibility to the community as a whole. With an eye on livelihood and environmental sustainability, the villagers have developed set limits on the amount of land available for each family, thereby limiting the amount of overall community land that will be converted to cashew nut production.

EXPERIENCES WITH PARTICIPATORY LAND USE PLANNING

Participatory land use planning (PLUP) has been utilized in many parts of Cambodia to protect the land and forest rights of forest-dependent and indigenous communities, and to guide the development of sustainable natural resource management systems. PLUP involves a series of community discussions regarding current and future land use practices and needs, combined with a mapping process that delineates village boundaries, protected forests, land for agriculture, and other use features. The three-village study found that communities like Tuy and Leu Khun, where PLUP processes were rushed or delayed, were not able to protect forests or village family land rights.

By contrast, the most successful PLUP activities occurred in Krala, where villagers began land use planning activities in 1998 with support from a German Technical Assistance (GTZ) technician. Initially, this process relied on "sketch" mapping that was used to develop a land use strategy and to serve as an advocacy tool when dealing with the provincial and national governments.[8] A geographic information system (GIS) map of Krala was completed in 2003 with support from an NGO, the Non-Timber Forest Products (NTFP) Organization. The NTFP Organization helped communities to produce PLUP maps that demarcated areas they had allocated for forestry and agricultural use, as well as protection (see figure 1). The organization went further to assist the communities in developing land use regulations and bylaws to manage their forest and land.

Villagers concerned about land alienation occurring in neighboring communities were quick to see the usefulness of mapping. As one villager noted, "If we have no map, land disputes will increase."[9] According to one study, 80 percent of the villagers interviewed said they wanted maps to display in their village.

> People feel that maps help them to stop illegal logging and other activities. They use maps as documents that establish their territorial claims with outsiders, such as government authorities and company representatives. Previously, villagers did not have clearly demarcated boundaries, and villagers would frequently cross each other's territories to make new swidden fields and

to gather non-timber forest products. During that time they also had disputes over benefits, but these were solved following accepted traditional procedures.[10]

In Krala, both sketch maps and GIS maps were created by villagers with help from the NTFP Organization. Sketch maps helped the community to understand its boundaries and to discuss land use zoning. GIS maps were created to provide local and national government with precise coordinates regarding boundaries. The tribal chief noted:

> Both maps are very good, and I need to display both of them in my village. I can remember the sketch map in my brain, and the GIS map has many signs, colors, and marks on it, and no one can understand it completely except clever people.[11]

While the NTFP Organization held several mapping training sessions, these largely involved members of the natural resource management (NRM) committee. The eleven members of the NRM committee in Krala Village met weekly to develop their plans and mapping activities. While the planning and preparations were performed by committee members, the final decisions on all matters that impacted the village were made at village-wide meetings, at which at least 90 percent of community members were usually present.

According to one NTFP Organization staff person, the people in Krala have worked hard on developing their land use plan, as well as their rules and regulations, which have been recently revised and are currently awaiting approval at the commune council level. Many people have dedicated a significant amount of time to attending meetings and collecting GPS data in the field. Informants shared that they are tired, and some members would like to take a break, but they all expressed enthusiasm for the work they are doing and pride in their accomplishments.

The community's rules and regulations set penalties and fines for offenses, such as causing a fire in the spirit forest or burial forest or for burning another person's field. However, no limits were set on the collection of non-timber forest products for subsistence use. If someone wanted to create a business that relied on the collection and sale of non-timber

Figure 1. Krala Village Participatory Land Use Planning Map, 2006

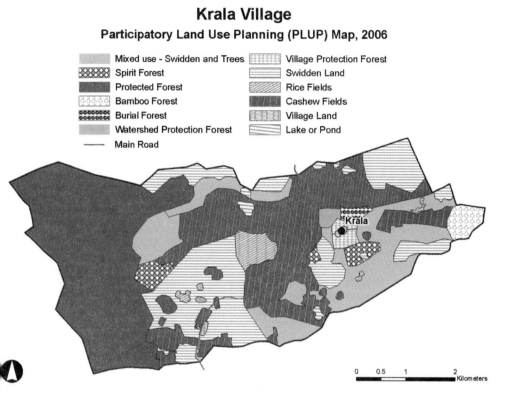

Krala Village
Participatory Land Use Planning (PLUP) Map, 2006

Mixed use - Swidden and Trees	Village Protection Forest
Spirit Forest	Swidden Land
Protected Forest	Rice Fields
Bamboo Forest	Cashew Fields
Burial Forest	Village Land
Watershed Protection Forest	Lake or Pond
—— Main Road	

forest products, they would be required to obtain permission from the community and possibly pay a fee to the community.

The village's rules and regulations allow land sales; however, members are only allowed to sell land to other members of the community. Selling land to outsiders is strictly forbidden, and would result in expulsion from the community. Krala, with the support of the NTFP Organization, has also submitted an application for a communal land title to the provincial authorities. The application has the support of the Ministry of Interior, but has not been approved by the Ministry of Land.

The land use planning and dialogue process in Krala was successful in guiding land cover change. For example, virtually all of the protected forest, bamboo forests, burial forest, and watershed and village protection forest remained under forest cover between 1989 and 2006. Forestlands that have been cleared and converted for cashew cultivation were all drawn from the mixed-use and swidden agricultural land pool designated in the plan.

FUTURE DIRECTIONS

Throughout mainland Southeast Asia, upland watersheds are being transformed as roads create market access, and traditional forms of agriculture and forests are replaced by commercial plantations and conservation areas. The transformation of land use and land tenure in the region raises numerous questions regarding their impact on the local culture, economy, and environment. The extent to which these upland transitions can be guided in ways that are supportive of social equity and environmental service goals depends to a great extent on the effectiveness of land use policy formulation and implementation.

LAND USE CHANGE SCENARIOS FOR RATANAKIRI

Lebel[12] describes four scenarios for the future of upland watersheds in mainland Southeast Asia. We have adapted these for this discussion (figure 2). The following pages describe the following four land use scenarios: (1) plantation economy, (2) parks and conservation, (3) traditional agro-ecosystem, and (4) diverse agro-ecosystem.

PLANTATION ECONOMY

In this scenario, economic growth is led by agricultural businesses. While farmers are interested in planting a number of market crops, rubber (*Hevea brasiliensis*) has emerged as the major commercial crop to replace traditional agriculture and secondary forests in Southeast Asia, a direct result of strong market demand from China, the world's largest consumer. In Cambodia, the Ministry of Agriculture, Forests, and Fisheries (MAFF) plans to expand the area under rubber cultivation from 50,000 hectares to as much as 800,000 hectares by 2015 (or 4.4 percent of the total land area of Cambodia). One of the implications of this scenario may be that the state will introduce policies to facilitate profit-making by agricultural businesses, which may or may not include smallholders.

Figure 2. Scenarios of Future Land Use Changes in Ratanakiri

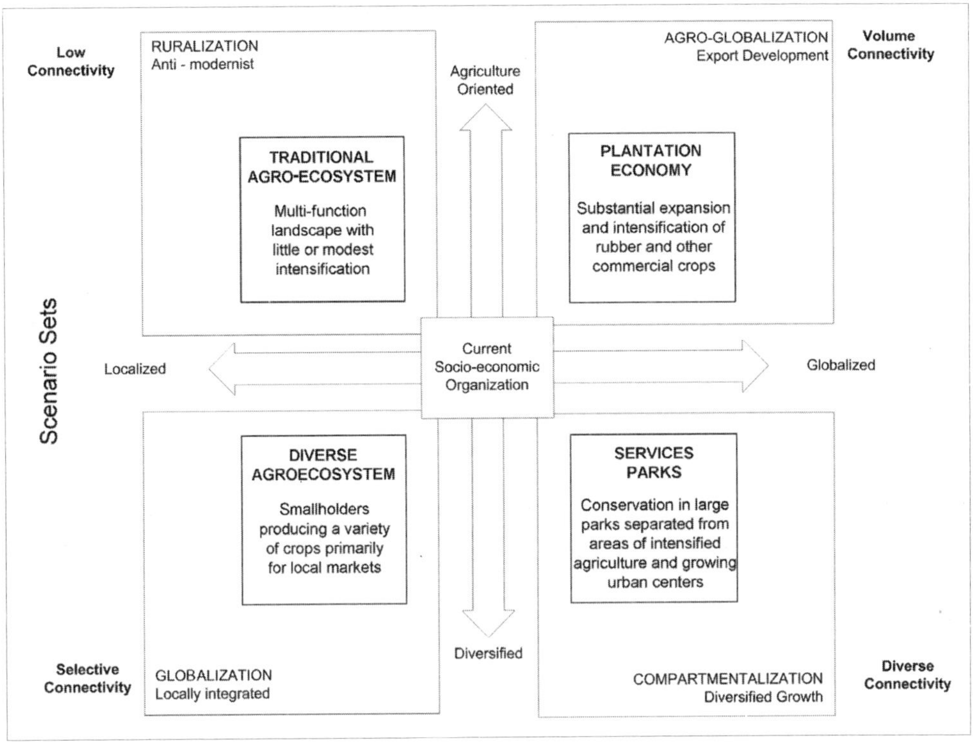

The issuance of economic land concessions (ELCs) could result in the transfer of land and forest resources previously utilized by local communities to external investors.[13] Over time, property rights may become predominantly vested in private firms; as ethnic minorities lose ownership of their land, they may become a source of flexible, low-cost, and mobile labor. A plantation economy would drive investments in road, transport, and communications infrastructure. But the large-scale expansion of monoculture cropping could lead to increased susceptibility to pest and disease outbreaks. Farmers' livelihoods would become increasingly vulnerable to changes in the market, climate, and other variables.

PARKS AND CONSERVATION

In this scenario, economic growth might unfold primarily through tourism that places a high value on forests, wetlands, rivers, and perhaps even "ethnic" diversity. Cambodia has already established Virachey National

Park (VNP) in Ratanakiri and Stung Treng Provinces. This scenario could result in the government acquisition of property rights from indigenous people and smallholders; segregation of areas of production and living from areas of nature and recreation; labor being drawn away to urban and agricultural areas; and investments in transportation and communication for tourism. A park scenario could also be envisioned that leaves local people on their land to practice traditional land use practices. The ability of local people to control their own lives, however, might be constrained.

Integrating communities into national protected area systems could also be explored through the establishment and recognition of community conservation areas (CCAs), known in Cambodia as community protected areas (CPAs). Many forest-dependent communities have traditions of forest protection, including sacred groves, burial forests, water source forests, and shelter forests, which could be strengthened through government policies and programs. Nonetheless, even formal protected areas like Virachey are vulnerable to alternative uses. Recently, mineral exploration rights were leased to private firms for much of the park. In coming decades, as economic growth accelerates in Cambodia, it will take substantial political will to retain areas designated for conservation.

Conservation initiatives can play important roles in either supporting indigenous stewardship and forest protection systems or in displacing them. Much depends on the conservation strategy that is adopted by government and implementing agencies. Communities can play highly strategic roles in controlling important drivers of deforestation, especially if their resource rights are recognized under conservation program agreements.

TRADITIONAL AGRO-ECOSYSTEM

For centuries, Ratanakiri, like much of the uplands of mainland Southeast Asia, has been managed by indigenous hill tribes who practiced a variety of forms of long rotation agriculture, supplemented with forest gathering and hunting. These systems are being replaced as the need for cash by upland communities is guiding them into more sedentary cash crops, while growing population pressures and land prices limited the availability of fallowed forests needed to support swidden farming rotations. Given current political and economic trends, traditional agro-ecosystems will likely continue to disappear unless there is a progressive lowering of private and public investments in regional infrastructure, either because these funds are targeted elsewhere or because of a prolonged global recession.

This scenario could reflect anti-globalization movements, dwindling agricultural trade, and an expansion of local exchange systems. Some of the implications of this scenario could be food security problems and land conflicts, as well as the rediscovery of local knowledge and appropriate technologies. It could also portend a rural bias in state policies. Rising costs for petrochemical-based fertilizers and pesticides, in addition to increasing fuel and transportation costs, may also create financial incentives for local communities to retain more traditional, self-sufficient farming systems.

DIVERSE AGRO-ECOSYSTEM

Under this scenario, significant but diversified economic growth occurs, but it would draw on local comparative advantages in agriculture, tourism, and perhaps mining rather than on the adoption of more uniform technologies and production systems. This scenario would require government policies that recognize the rights of minority peoples, integration of the national park into the wider landscape, and development of transportation and communication systems sufficient to meet the needs of the crops grown and of agro-ecological tourism. This would appear to be the most "ideal" of the scenarios outlined here in terms of protecting local resource rights and sustainable development. Farmers' livelihoods would be based on secure tenure rights, as well as diversified agricultural production for local and international markets.

CONCLUSION

As land is increasingly viewed as a marketable commodity, especially if planted with valuable crops like cashew or rubber, economic incentives are created to develop forestlands for income or for sale. Land as a market commodity conflicts with indigenous land management concepts that view land and forests as a communal resource to be kept intact for future generations. Demographic growth, both through natural increase and immigration, combined with corruption and economic expansion, challenges the viability of more traditional land management models as natural resources become scarce. Nonetheless, there are many indigenous communities around the world profitably and sustainably managing extensive areas of land held under communal tenure, while providing significant national benefits and environmental services. Indigenous management and communal ownership do not necessarily conflict with modern commercial agricultural or forestry production systems, and can in fact enhance agricultural transitions and sustainability.

Indigenous communities in Ratanakiri are gradually shifting their land use systems to reflect a greater emphasis on cash crops, especially cashews. In Krala, by retaining communal ownership, the village has been able to accelerate cashew planting when compared to neighboring communities and has achieved a high level of equity in the process. An effective process of land use planning ensured that the Krala community retained conservation forests, as well as a sufficient production forest reserve to allow for a continued swidden and cashew expansion area. This experience stands in sharp contrast with Tuy, where land tenure insecurity and poverty have led to a rapid process of land alienation, forcing local villages to open new fields in remote forests once considered "protected areas." In Tuy and Leu Khun, the formulation of participatory land use planning (PLUP) documents has had limited effectiveness in guiding land use change and protecting indigenous land rights, though in the case of Krala, it is apparent that these tools can be effective if based on extended community capacity building.

There are, however, new legal tools that can be used to stabilize the tenure situation in Ratanakiri and limit illegal land sales. Over the past decade, the Royal Government of Cambodia has drafted a new Land Law, Forest Law, and Community Forestry Sub-Decree, all of which have elements that could help ensure more equitable and sustainable use of its natural resources. Unfortunately, the lack of financial and human resources and the constraints of competing policy and political agendas have resulted in these policies and legislation not being effectively operationalized and implemented. The RGC, with support from the donor community, needs to move forward proactively to utilize existing legal mechanisms to support indigenous community efforts to stabilize forest and land resources over the coming decades.

It is clear that land use is changing rapidly in all three study villages, reflecting broader patterns operating in Ratanakiri and other parts of the uplands of mainland Southeast Asia. Some of this change reflects a broad-based agricultural transition from forms of swidden farming to commercial cash cropping, especially the adoption of cashew trees. The commercialization of farming systems has created a new source of income for many indigenous families, while at the same time stimulating land markets and accelerating land alienation. Communities like Krala that have strengthened their indigenous institutions and established clear policies on land use and tenure are successfully building on new market opportunities, while also sustaining their forest resources and cultural institutions. By contrast, communities like Tuy are being rapidly transformed into areas where villagers sell their land and migrants move into the areas. At present, many Ratanakiri villages are like Leu Khun, struggling to maintain community lands and forests in the face of growing pressures. Whether these communities will share the fate of Tuy and experience a chaotic pattern of land use and tenure change, or stabilize their resources like Krala and systematically move into new modes of production, depends on a number of factors. Even Krala may succumb to disintegration if social systems are not respected.

A key variable is the extent to which these communities will receive support from outside agencies, including both NGOs and government programs, and protection from illegal land speculators. In all study areas, villagers noted the importance of NGOs in helping them to retain their communal land and learn how to deal with local government and market forces. The study also showed that long-term, sustained community building is a key to success in establishing viable community institutions that can guide land use and tenure policymaking.

Local NGOs have made a tremendous contribution to indigenous communities in Ratanakiri, but their capacity needs to be increased, both in skills and coverage. They also need to find ways to integrate further with indigenous institutions to gain a greater degree of community ownership. Community networks that are less dependent on nonindigenous structures need to be fostered to further strengthen civil society institutions in Ratanakiri.

Finally, local government officials and community leaders require training and guidance in national land policy, as well as an open and transparent framework for dialogue at the commune, district, and provincial level. There is an urgent need to clarify land and forest resource management rights and responsibilities throughout the province, especially in an effort to protect the ancestral domain claims of the region's indigenous communities. The FA has the role and responsibility to demarcate the state public forest domain and to determine which areas are suited to FA-recognized community forestry. The FA also has the role of coordinating with the Ministry of Land Management, Urban Planning and Construction, in order to delineate land for inclusion in communal titles of indigenous communities. While much of the legal framework is in place to begin establishing recognized community forestry sites and issuing communal titles, the priority must be placed on the mobilization and strengthening of communities.

Economic land concessions need to be kept out of indigenous peoples' areas, as they are creating conflict and causing displacement, while not providing effective stimulants for economic growth. As a recent United Nations report concludes:

> Economic land concessions have not led to increased agricultural productivity or economic growth in Cambodia, and large areas of conceded land have been left idle or underutilized. As recommended by the World Bank Poverty Assessment 2006, secure land title and family-based or smallholder agriculture would improve development outcomes for rural communities. Community-based initiatives for land and natural resource management should also be prioritized.[14]

The landscape of Ratanakiri is being transformed as forests are being cleared at an estimated annual rate of 5 percent along Highway 78 that runs from the Mekong River in the west to the Vietnamese border in the east. This study indicates that traditional communities in Ratanakiri have lost nearly 40 percent of their forests over the past sixteen years. In Tuy, at the current rate of clearing, all village forests will be gone by 2018. Land purchases by outside investors, mostly illegal, are rapidly displacing local families, who are driven further into forests once zoned for conservation in order to create new agricultural lands. Community respondents in Tuy estimated that nearly 80 percent of households have already sold their land to migrants and investors, many times with coercion being a significant factor. As forests are cleared, the land is being replanted with rubber and cashews. While the processes of forest conversion to estate crops and indigenous land alienation are most advanced in Tuy, other communities are following these patterns, which are spreading throughout much of the province.

The problems presented by this rapid change relate to issues of social justice and continuity, economic equity, and the provision of environmental services. Land alienation is a classic problem for indigenous peoples. Loss of ancestral domain leads to an erosion of cultural identity, especially for cultures that hold their land and forest resources communally, as do the peoples of Ratanakiri. Illegal land purchases and the leasing of large economic concessions are increasing rapidly, often facilitated by local government officials in exchange for commissions. Disempowered and impoverished communities urgently require implementation of both policies and programs that protect them from exploitation. This needs to start at the most basic level, with the implementation of the Cambodian Constitution and the 2001 Land Law, before new policies and laws are developed.[15]

The case of Krala demonstrates that supportive measures by government and civil society organizations can create environments where indigenous peoples can retain their cultural identity while successfully participating in a market economy, supporting national economic development goals, and implementing sustainable agricultural transitions. Community management needs to be strengthened to deal with increased conflicts over land and forests. In Ratanakiri, the traditional decision-making unit for governance and conflict resolution is at the village level. The RGC already has the necessary legal and policy framework in place to protect forest-dependent peoples' resource rights. Now the government must demonstrate the political will to actively enforce and implement this framework. The people of Ratanakiri possess an in-depth knowledge of their environment that has tremendous potential value for informing management decisions, as well as for playing an active role in Cambodia's economic growth.

NOTES

1. C. Meyer, "Les Nouvelles Provinces: Ratanakiri – Mondulkiri." *Revue Monde en Developement* 28 (1979): 682–90.

2. Bruce McKenney, Yim Chea, Prom Tola, and Tom Evans. "Focusing on Cambodia's High Value Forests: Livelihoods and Management" (Phnom Penh: World Conservation Society, November 2004), 15.

3. Graeme Brown, Jeremy Ironside, Mark Poffenberger, and Alistair Stephens, "Formalizing Community Forestry in Ratanakiri Province," in *Cambodia: Linking Indigenous Resource Systems to Government Policies and Programs* (CFI: Phnom Penh, Cambodia, 2007).

4. Jeremy Ironside and Ian G. Baird, "Wilderness and Cultural Landscape: Settlement, Agriculture, and Land and Resource Tenure adjacent to Virachey National Park, Northeast Cambodia" (Biodiversity and Protected Area Management Project – BPAMP, DNCP/MOE: Cambodia, 2003), 25.

5. "Ratanakiri Provincial Development Plan 2006–2010" (Provincial Department of Planning, Ban Lung, Ratanakiri, 2005).

6. Ibid.

7. Ian G. Baird, *The Ethnoecology, Land-Use, and Livelihoods of the Brao-Kavet Indigenous Peoples in Kok Lak Commune, Voen Say District, Ratanakiri Province, Northeast Cambodia* (Ban Lung: NTFP, 2000), 20–30.

8. Prom Meta and Jeremy Ironside, "Effective Maps for Planning Sustainable Land Use and Livelihood," in *Mapping Communities: Ethnics, Values, Practice*, ed. Jefferson Fox, Krisnawati Suryata, and Peter Hershock (Honolulu, Hawaii: East-West Center, 2005), 29–40.

9. Ibid., 33.

10. Ibid., 34.

11. Ibid., 36.

12. L. Lebel, "Multi-level Scenarios for Exploring Alternative Future for Upper Tributary Watersheds in Mainland Southeast Asia," *Mountain Research and Development* 26, no. 3 (2006): 263–73.

13. Cambodia Office of the High Commissioner for Human Rights, "Economic Land Concessions in Cambodia: A Human Rights Perspective" (Phnom Penh: United Nations, 2007).

14. Ibid., 22.

15. In India, during the colonial and independence era, national legislators and policy makers adopted laws to prohibit the sale of land in areas designated as scheduled tribal areas in regions like northeast India and the central tribal belt.

CF Training program at Prasat Sambo, Kg. Thom

CF members on patrol

CF educational outreach-shadow puppet theatre

Tree planting ceremony, Mondulkiri

Bun Suluth, HE Ty Sokhun at Oddar Meanchey REDD project site

Samraong Pagoda monks discuss plans for forest protection

Fire is an annual threat in the dry season

Economic land concessions clear and burn forests for agriculture

Evergreen forest protected by CF groups

Dry deciduous forest in Oddar Meanchey

onle Sap flood forest protected by CF group in Kg. Phluk

Monk patrolling forest
Oddar Meanche

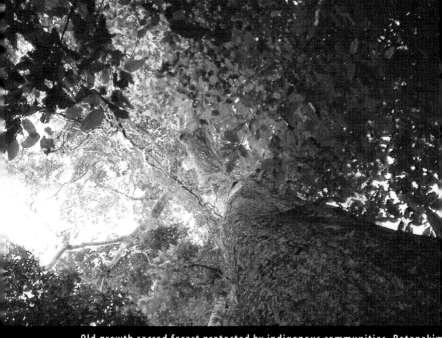

Old growth sacred forest protected by indigenous communities, Ratanaki

Village children gathering firewood

Villager gathering wild forest honey, Sre Ambel

Ethnic minority children, Yeak Lom

Kg. Phluk village in flood season, Tonle Sap

Hut, Ya Poey, Ratanakiri

Khmer grandmother and child

PART III

THE VALUE OF FOREST
MANAGEMENT TRADITIONS

CHAPTER 6

STRENGTHENING FOREST CONSERVATION
THROUGH THE BUDDHIST SANGHA

CHANTAL ELKIN

THIS CASE STUDY DOCUMENTS THE EXPERIENCE OF A GROUP
of Buddhist monks who have pioneered the establishment of one of
Cambodia's largest community conservation forests. This story discusses
the motivations, initial activities, challenges, and eventual success in
obtaining national recognition for the "Monks Community Forest." The
chapter also describes the national and local contexts in which this grassroots conservation initiative emerged, and how it is saving some of the
last lowland evergreen forests in mainland Southeast Asia. It ends with a
discussion of the potential for engaged Buddhist conservation activities in
Cambodia.

The ongoing destruction of natural resources in Cambodia has
attracted a range of international conservation organizations and, with
them, large amounts of funding aimed at safeguarding areas and species
of global and national biodiversity importance. These NGOs operate
from a set of Western values and beliefs and commonly focus on solving
the problems of natural resource destruction at the symptomatic level,
using a variety of financial and other incentives and deterrents. While this
Western approach has its merits and successes, at times it meets resistance
from both the government and the broader population and sometimes

fails to achieve real and lasting buy-in. This case study demonstrates that an alternative model exists that can fundamentally shift the views Cambodians hold about their natural environment—one that springs from the traditional values and customs of Khmer culture.

MONKS COMMUNITY FOREST

In recent years, individual Buddhist monks have been initiating and guiding grassroots environmental protection in their communities, ranging from tree planting to safeguarding forests to environmental education. These projects are demonstrating the transformational power of Buddhism in shifting attitudes towards the environment. One impressive example is the 18,261-hectare Monks Community Forest (MCF) in Oddar Meanchey Province, northwest Cambodia. On their own initiative, the monks of the Samraong Pagoda have acquired formal, legal management of what is now Cambodia's largest community forest. They have established forest patrols and organized community volunteers, demarcated forest boundaries, raised environmental awareness among local communities, attained the support of government authorities and NGOs, and by all accounts have significantly reduced forest crime in the MCF. Many local people interviewed believe that the MCF is now one of the best-protected forests in the province.

ORIGINS OF THE MCF

At the helm of the MCF is Venerable Bun Saluth. In 2001, this young, charismatic monk returned to his home province of Oddar Meanchey after five years in a monastery in northern Thailand. While there, he was deeply impressed by the conservation ethic of the Thai monks, who planted trees as part of the pagoda's efforts to counteract the severe flooding and drought that had followed years of destructive logging.

Upon his return to Oddar Meanchey, Ven. Saluth was shocked to see huge swathes of forest gone. Logging had been rampant in the lawless post-war years of the late 1990s.[1] Oddar Meanchey's high-value timber had been heavily extracted during this period by Thai companies, in collusion with the Cambodian armed forces and government authorities, often backed by Cambodian businessmen.[2] From 2002 to 2006, the province's average annual deforestation rate was 2.1 percent[3]—versus a national rate of 0.5 percent—from ongoing logging, encroachment, forest fires, and economic land concessions.[4]

Ven. Saluth was inspired to take action. In late 2001, an opportunity presented itself. A fellow monk who regularly meditated in the forest told him of a near-intact expanse of forest located in Samraong District. Ven. Saluth wrote a letter to the provincial governor asking permission for the monks of his pagoda to oversee protection of this 3,000-hectare area. By this point, he had become head monk of the Samraong Pagoda and deputy chief monk of Samraong District, making him one of the highest-ranking monks in the province.[5]

The governor asked Ven. Saluth to expand his request and on February 7, 2002, wrote a letter granting Ven. Saluth the authority to protect 18,261 hectares of evergreen and semi-evergreen forest. The area, known in Khmer as Sorng Rokavorn, was then called the Monks Forest, and is now called the Monks Community Forest, or MCF.

Forest protection throughout the country was very weak at this time. The Forestry Law had only been enacted in 2002, and many areas were still subject to anarchic logging. Once permission was granted, it was Ven. Saluth who established the rules governing the MCF: logging and hunting were now forbidden, but locals could harvest non-timber forest products (NTFPs) and fish using traditional methods.

Ven. Saluth enlisted several monks from the Samraong Pagoda to begin protection efforts by erecting boundary demarcation poles around the perimeter of the protected forest and by patrolling the forest on foot, on the lookout for violations of the new regulations. Slowly, they began to raise awareness among local people about the new protected forest, and in 2002 the monks began working in earnest with two nearby villages. In 2004, more villagers stepped forward, now convinced that protecting the MCF would also be of benefit to their communities. More and more village volunteers joined the protection efforts when they realized, as time went on, that the monks were protecting the area for the good of the community, while elsewhere in the province forests were fast disappearing.

PRESSURES ON THE FOREST

From the beginning, it was clear that illegal logging was the greatest threat to the MCF. Oddar Meanchey Province was a Khmer Rouge stronghold until about 1998, when the last peace agreements were settled with the government. Many of the people living there were still armed even after the war, and poverty was endemic. It is not surprising that in the early years, the monks were at times confronted with armed, angry loggers, some of whom even threatened the lives of several monks. In addition, because at least one military base was located near the forest boundary, soldiers would frequently enter the MCF to fish, hunt, and log.

Economic land concessions, a ubiquitous problem throughout the country, are an added pressure. The government typically awards concession rights to companies, which then clear forested land of trees. The land is used for plantations, mining, or other resource extraction activities. The government usually does not recognize the land rights of rural people, who in many cases occupied and farmed the land for years. In Cambodia, large companies often win out over the small landholder, and poor rural people—the majority of the population—are further marginalized, become landless, and then look to the forest to clear new land.[6]

FOREST PROTECTION

The MCF monks have developed what is widely viewed as an effective and unique system of forest protection based on Buddhist principles. The involvement of monks in every aspect of protection is the MCF's most distinctive feature. The monks not only initiated forest protection and manage all operations, but also participate in forest demarcation, community outreach, and, most notably, patrols. Almost all of the people interviewed believed the involvement of the monks in the management and patrolling of the MCF to be the main reason for its effective protection, particularly when compared to other community forests in the province. This discussion will highlight a few of the unique aspects of the MCF protection system.

ORGANIZATIONAL STRUCTURE

Initially, the monks operated alone, identifying the conservation area, seeking formal protection status, designing the regulations, erecting demarcation signs, and carrying out patrols. Before long, they began to reach out to villagers living near the MCF, raising awareness about the protection efforts and asking for volunteers for patrols. By 2006, about twenty people from eight villages were volunteering their time to patrol with the monks, and as of December 2008, there were forty volunteers from six villages helping to protect the MCF. The villages are all located on the outskirts of the MCF, and only four indigenous families live inside the boundaries of the protected forest.

The forty volunteers sit on "village subcommittees," each with five to seven members. The members are responsible for patrolling the MCF and for liaising with the community forestry management committee (CFMC), the main body overseeing the MCF operations. The CFMC meets regularly at the Samraong Pagoda in Samraong town, which is over an hour's drive from the protected forest. It is responsible for developing the operational strategy, overseeing the subcommittees, organizing the budget, and taking care of all administration. It also acts as a liaison with local, provincial, and national authorities and with donors and NGOs. Ten people sit on the CFMC, including three monks and seven laymen, three of whom were formerly monks. These are all male, but Ven. Saluth would like to add a few women on the committee to help with administration.

RULES AND REGULATIONS

The MCF regulations prohibit all logging and all hunting of animals, regardless of their conservation status, reflecting the Buddhist aspiration to ease all suffering of living beings, and to live by virtues such as "doing no harm" and compassion. As such, the regulations put in place by the monks are much stricter than the Cambodian Forestry Law, which categorizes species according to their threatened status, and permits many common species to be hunted for subsistence use within the country's protected forests. The MCF rules do, however, allow fishing by traditional methods, and collecting and selling NTFPs such as mushrooms, resin, and bamboo. The monks have also shown leniency with the rules in permitting some farmers to use old paddy fields inside the MCF, based on an agreement that the farmers will not clear new land. In some cases, needy families can harvest some timber in the MCF if they obtain specific permission from the monks.

ORDAINED FOREST

One of the most interesting features of the MCF, and one of the most effective in deterring illegal activity, is the "ordained forest," situated at the northern end of the MCF. Each year monks conduct a tree ordination ceremony in which they bless an area that includes about two hundred of the MCF's largest and oldest trees, wrapping their trunks in saffron robes to sanctify them. To cut down a tree or hunt wildlife within the ordained forest is considered as serious as harming a monk, and is viewed as an act of demerit with negative repercussions for the next rebirth.[7] There is also a belief that animistic spirits are often found in the largest trees. Since the first ceremony, they have not had any incidents of illegal activity in the ordained forest.

PATROLS

At the heart of the MCF protection system is the regular patrolling of the forest to stop offenders from carrying out activities in violation of the regulations. These patrol teams are made up of unarmed monks and villagers. A core group of forty village patrollers and CFMC members, as well as volunteers from the Samraong Pagoda, participate in the patrols when they have time. The participation of monks in forest patrols is highly unusual in Cambodia, and stands in stark contrast to the traditional pagoda-based life of a monastic. Their presence on patrol is by all accounts an effective deterrent to offenders, more so than if the patrols were made up solely of village volunteers. This is likely due to their respected status in society and the belief that harming a monk brings negative karmic retribution.

In addition, Ven. Saluth's position as a high-ranking monk and his strong relationship with provincial authorities usually ensure that the soldiers leave the forest without incident.

When patrolling, the teams drive as far as they can into the forest on motorbikes, then leave the bikes and walk around the forest listening for chainsaws and looking for intruders. When there are signs of illegal activity, they quietly investigate and determine whether the offenders are armed. In the vast majority of cases, if the patrols come across people who are logging or hunting illegally, they stop them, confiscate their equipment and materials, and ask them to sign a contract stating that they will not commit a second offense in the MCF, using their fingerprints as signatures. Each patrol carries a mission letter, a form for recording the details of the event, and a contract. If they suspect the intruders have guns, they report to the authorities (FA/police/military police) and ask them for assistance. More often than not, however, by the time they arrive, the offenders will have usually gone, as it can take several hours or up to a day for the authorities to travel to the MCF.

The involvement of the village volunteers on patrols has been critical to the protection of the MCF, greatly increasing the presence of monitors inside the forest. As local support has increased for the MCF, villagers increasingly report sightings of illegal activity to the patrol teams, resulting in an excellent system of intelligence gathering and reporting to support law enforcement efforts. Ven. Saluth claims that illegal loggers rarely escape now because the villagers work so well with the monks, sometimes even surrounding the offenders if they attempt to flee.

INTERACTION WITH OFFENDERS

An unusual aspect of the MCF operating protocol is giving the offender three chances before serious punitive action is taken. This is in line with the Buddhist way of teaching, says Ven. Saluth: People are given a chance when they make a mistake. But if someone is caught a third time, they will be arrested and sent to court, although this has yet to happen.

Another of the guiding principles on patrol is interacting with offenders in as peaceful a way as possible. As already noted, the monk and villager patrols are unarmed. It is also important to the monks that when a patroller meets an offender in the forest, harsh and angry words are not used. The village volunteers said that they follow the monks' example and try to do the same. One of the other nonviolent and effective methods of dealing with offenders is taking their photograph, which normally causes them to retreat from the forest, frightened that their photo will be shown to the authorities. If there is a particularly serious threat or the offenders are armed, a large group of monks will go to the forest together, which both intimidates the offenders and protects the monks. These approaches seem to work well as nonviolent deterrents.

In summary, the monks have developed a "soft" approach to law enforcement, which reflects their underlying desire to cause as little suffering as possible to the people they encounter. The system includes unarmed patrols, treating offenders without anger, a three-strikes-you're-out policy, the setting aside of an ordained forest, and the use of photography and large groups of monks on patrol to deter offenders. The very involvement of monks in the management and patrolling of the MCF brings legitimacy to forest protection and acts as a powerful deterrent to forest crime. These efforts all reflect the monks' underlying motivations to abide by Buddhist principles while safeguarding the forest.

RELATIONS WITH KEY STAKEHOLDERS

As with many engaged Buddhist movements, the MCF has a charismatic leader at its helm. It is a testament to Ven. Saluth's leadership abilities that so many different stakeholders now contribute to the MCF's protection, including national, provincial, and local government; local communities; NGOs; and the Samraong monastic community.

Ven. Saluth seems to have a gift for establishing good relations with authorities at all levels, and has the confidence and the standing to call the head of the FA, the provincial governor, and other influential people when he needs further support. Senior government officials have visited the MCF and, in doing so, have strengthened the legitimacy of the protected forests in the eyes of local and provincial government officials, villagers, military, and other stakeholders and potential offenders.

The Samraong monks have also developed broad support among communities living near the MCF, but this has been a gradual process acquired through considerable outreach. There has been tension over the years with several of the communities that see the MCF rules as limiting their access to wood and farmland. In some villages, people lamented that they did not have enough land to farm, largely because land was claimed either by the MCF or by companies for plantations, although there was a feeling that the land would at least be protected under the MCF's jurisdiction.

Three villages in particular, located in the southern part of the MCF, have been less than cooperative. The monks suspect the village chiefs, backed by powerful officials and businessmen, of hiring people to clear land in the MCF to sell as concessions. Ven. Saluth also acknowledged that outreach in these villages had been weak. In late December 2008, however, the monks met with these southern villages to raise awareness about the benefits of the MCF and to organize elections for their subcommittees. As a result of these activities and the legalization of the MCF's community forest status, relations have improved immensely, and these villages now officially form part of the MCF.

Although they are prohibited from cutting down trees or hunting inside the protected forest, villagers are allowed to harvest NTFPs like bamboo, fruit, and mushrooms. Mushrooms are particularly lucrative. Three varieties, in particular, are highly sought after, each with its own season. They can fetch $150–$200 for about fifteen days of harvesting, representing an important source of income. Villagers also used to burn trees for resin, but this has stopped. Access to NTFPs and further awareness-raising efforts appear to have gone a long way in building trust between the monks and local villagers. Many of the villagers confided that they see how the monks are protecting the forest for the benefit of current and future generations and that they are trying to follow their example.

LIMITATIONS

The Samraong monks have done an impressive job protecting the MCF, but they are aware that they have certain limitations. Perhaps the greatest challenge is meeting the costs of protection. The monks have little experience with international donors and fundraising. Financial support for one patrol team, which is made up of four people patrolling for about fifteen days a month, costs an average of $4,000 per year, not including repairs or extra costs that may arise.

The MCF has been largely financially self-sufficient, and much of the costs of protection come from donations to the pagoda. The monks and volunteers also meet a significant portion of the costs from their own resources and only occasionally receive money from the management committee (CFMC). Like the monks, the forty village subcommittee members are unpaid. One volunteer reported that he spends about $20 of his own money per month on petrol. Another volunteer said he receives money from the main committee perhaps one in ten times. In addition, the volunteers use their own transportation and phones. This is a significant expenditure given the poverty of the area, in a country with one of the lowest per capita incomes in the world, just over $700 in 2008.[8]

Almost everyone interviewed felt there were not enough patrols for such a large area. The monks and villagers also expressed the desire for greater support from local law enforcement authorities for cases involving armed offenders, to make arrests, and to send cases to court. The monks currently do not have the funds needed to pay the FA for their assistance. The patrols also require more equipment and supplies. They would like to build several new patrol stations in the south where the main threats persist.

Many people interviewed expressed their concern that volunteers are neglecting some of their livelihood activities in order to patrol, and that it is necessary to compensate them for their work. During rice growing season, the volunteers leave their wives to tend the fields, making it more difficult for families to meet their basic needs. The monks would like to offer salaries for all members of the project. It would also be beneficial to have some posters, signs, and banners in the villages and on the road to the forest, in order to raise awareness about the location and regulations of the MCF, and to make sure people understand they will be penalized if they violate the regulations.

Finally, the monks would like to improve their protection efforts through training courses for all members, and to participate in meetings related to forest protection, in order to improve their understanding of forest conservation in Cambodia. They would also greatly benefit from training by forest experts in best practices of forest protection, and by enforcement officers who could help them design effective patrol plans.

The MCF has received some technical and financial assistance, but the team expressed a desire for more external support for the activities listed above. At one time, the monks received a bit of support for travel, materials, and training from the local NGO, Buddhism for Development, which continues to give some technical support when the monks request it. Community Forestry International (CFI) was instrumental in providing core support to Ven. Saluth when he began to organize his strategy for forest protection. Later, together with the Forestry Administration, CFI provided a good deal of technical and financial help to enable the MCF to acquire community forestry status and to meet the guidelines of the Community Forestry Law. CFI also provided funds and technical training for border demarcation, awareness-raising, and patrols.

In 2009, the Alliance of Religions and Conservation (ARC), together with the Interfaith Power and Light organization, raised $10,000 to support patrol work and community livelihood assistance. In 2010, Ven. Saluth won special recognition at the prestigious UN-sponsored Equator Prize awards and, with it, $20,000. As a result, the MCF is receiving more international attention as an area of significance.

BUDDHIST BELIEFS AND ENVIRONMENTAL PROTECTION

It is often the case that engaged Buddhist movements are initiated by charismatic leaders. The MCF is no exception. Many of the monks and community members involved in the MCF said they had joined the protection efforts because of their relationship with and regard for Ven. Saluth. The fact that an entire pagoda of approximately forty monks is mobilized to protect the MCF under Ven. Saluth's guidance is testament to his leadership. On closer inspection, however, the Samraong monks expressed a more fundamental motivation for their participation in protection efforts: once the link was made between the Buddhist view and environmental protection, they came to see conservation activities as a natural extension of their Buddhist faith.

VIRTUES OF FOREST PROTECTION

The monks involved with the MCF see many Buddhist virtues as under-lying their forest protection activities, including, among others, not causing harm to any living creature (*ahimsā*), compassion (*karunā*), loving kindness (*mettā*), generosity (*dāna*), simple living (*arāga*), and wisdom (*paññā*). One of the most important of Buddhist virtues is *ahimsā*, often taken to mean to do no harm to any living being. The Samraong monks' interpretation of this teaching includes not only doing no harm, but *actively preventing* harm as a way of relieving the suffering of other sentient beings.[9] Because *ahimsā* was emphasized so often in the inter-views, it came across as the strongest impetus for the monks' conservation activities.

The monks see logging a tree or hunting an animal as taking a life, and not only counter to the virtue of *ahimsā*, but also a violation of the first of the Five Precepts, the basic duties or commitments undertaken by a monastic and the basis of Buddhist morality. They believe there can be no more moral action than safeguarding the forest and its creatures from the destruction occurring throughout Oddar Meanchey and across Cambodia, and from the ensuing poverty for farmers and other rural people who depend on the land.

The monks also interpret forest protection as part of the Buddhist teaching of dependent origination (*Paticcasamuppāda*), which sees all life as interrelated. The nationally revered monk Venerable Maha Goshananda put it this way: People and forests are connected. If we take care of ourselves we will take care of the forests, and the forests will care for us in return. Ven. Saluth asserted that a monk's duty is to encourage the virtue of simple living (*arāga*) and to root out the evil of greed in the mind, which can lead to environmental destruction. Monks need to act as guides, showing people the middle way, and demonstrating how with few material posses-sions they can survive and live peacefully. In short, the monks interviewed expressed the view that protecting the MCF was one of the ways that they could help alleviate the suffering of living beings, and in doing so, live the ethical life the Buddha encouraged in his followers.

KARMA

When people see monks they are happy, because they receive their blessings....When people see me, they kneel down. They respect me.[10]

In Cambodian society, monks are considered "fields of merit," with the ability to impart blessings on their community and to transfer good merit to laymen and their deceased ancestors. Accumulating good merit improves one's chances of a favorable rebirth.[11] Laymen can gain good merit by making offerings to monks, and this has become a central practice in Cambodian Buddhism.[12]

A basic belief that surfaced in the interviews is that protecting life, including trees and wildlife, results in good karmic retribution, and that causing harm incurs negative karmic retribution. One monk expressed that forest protection helps people and thereby gains him good merit. Some villagers view their work with the monks as an offering, and thus a way to accumulate good merit. One village patroller said, for example, that because he has no money to offer to the temple, volunteering with the monks to protect the MCF was his form of *dāna* (generosity, a Buddhist virtue).

Fear of incurring negative karma is also a motivating factor. One monk said that if someone kills a monk, they descend into "deep hell." A village volunteer insisted that if you don't respect monks you can develop bad karma, and another that hurting a monk is a sin because he is like the Buddha. The village volunteers also expressed a fundamental belief that "doing good things produces good things," and that destroying the forest reaped bad karma, while protecting it helps the community and produces good karma.

Among some villagers, however, was a sense of frustration, a feeling that at times the monks are too demanding. Clearly some monks are not afraid to use their status to encourage cooperation if they believe it is for the greater good. The Samraong monks recognize that they are influential and that they can benefit conservation in ways that other people cannot. Indeed, most of the monks and villagers voiced a concern that if only villagers are involved with protection efforts, the forest will certainly be lost. The monks play a critical role in stopping illegal activity when villagers are scared to intervene. In a country where most people are Buddhist and believe in karmic retribution and in the power of monks to contact the spirit world, the sight of a monk on patrol can be intimidating.

Another driving belief is the view of the continuity of karmic life, in which a human can be reborn as an animal, or had been an animal in a past life. Some scholars argue that this belief in a deep connection between all forms of life makes it more difficult to cause harm to any creature because it is as if the harm is being done to oneself.[13] The monks cited several parables about the Buddha's past incarnations as creatures of the forest. In the Jatakas, the 500 fables of the Buddha's past lives, each story conveys how to live a moral life, using trees, spirits, and animals as protagonists. The fables directly link Buddha with the forest, and as the scholar Sahni asserts, "make the boundaries between animals and humans much more fluid than in the West."[14]

CONNECTION TO THE LIFE OF THE BUDDHA

In response to the question "Why are you protecting the forest," Ven. Bun Saluth explained: *"The tree is a symbol of life, and sacred to Buddhists. Buddha was born under the tree, attained enlightenment under the tree, and died under the tree."* One monk said that they highlight this connection between nature and the life of the Buddha, believing that the villagers respond well to the motivation of Buddhism as justification for forest protection. They find this a more effective approach than scientific arguments that the villagers may not understand. He says:

> The one story that I always explain and tell them to protect the forest is mainly related to the Buddha's life… this is the house of the Buddha, it is the shelter of the Buddha, so if you are a Buddhist, you have to respect, to take care of the forest, because this is our father's house and we have to respect the father.

BELIEF IN THE SUPERNATURAL

A belief in spirits seems to be a secondary but added incentive to maintain the integrity of the forest. In addition to the belief that trees can be the abodes of Buddhist deities called *devas* that take care of the forest, Cambodian belief includes a range of ancestral and animistic spirits called *neak ta*, some that are specific to the forest and other natural areas. It is thought that big trees in particular are homes to these spirits and are under their protection. The monks told me that harm will befall anyone who destroys the homes of the *devas* and *neak ta*. Most of the monks also believe that animals have guardian spirits, including elf-like guardians called *mrenh kongveal*. At first glance, the belief in spirits may seem counter to strict Buddhist orthodoxy and practice. On closer inspection, it becomes clear that the Khmer religious belief system is a complex one consisting of tantric Theravada Buddhism, remnants of Brahmanism, and belief in a vast range of spirits. These seemingly contradictory beliefs function as one system and are the lens through which the Khmer see themselves and their world.

NATIONALISM/KHMER HERITAGE

Many of the responses, both from the monks and villagers, expressed a strong feeling that the forests are "our heritage, like Angkor Wat." Monks and villagers felt that it is the right of future generations to inherit healthy forests and ecosystems so that they can see the full range of Cambodia's wildlife and enjoy the forest's many benefits. The monks made a more explicit link than did the villagers, however, between forests, religion, and nationhood, and asserted that these three elements of Khmer society should all work together. They clearly feel that protecting the MCF is their patriotic duty, as well as their religious duty. "Monks should be involved in forest protection because it benefits the nation and is important for religion," asserted one monk. Such comments make sense in a country where the intertwining of Buddhism and nationalism has a long history. This is expressed in the country's motto of "Jati (Nation), Sasana (Religion), and Mohaksatra (King)."[15]

The above discussion has highlighted the role of belief as the primary motivator for forest protection among the monks and the Buddhist communities living near the MCF. The monks believe that through conservation they are walking the path of the Buddha, helping relieve the suffering of sentient beings, and protecting life; pursuing a life of virtue, while helping others live according to Buddhist principles; and accumulating good merit and helping others do the same in the hope that it will benefit them in future lives. The monks have also been able to cultivate a conservation ethic among villagers by explicitly linking nature conservation to the life of the Buddha, and to Buddhist principles such as karma and moral living. The fact that the monks are giving their time, resources, and efforts to forest protection sends a strong message to the community that to live well and morally, it is important to respect all life and to conserve the forests and natural resources.

RECENT PROGRESS

Despite the challenges of the first few years of protection, the MCF is now regarded by local villagers and authorities as one of the best-protected community forests in Oddar Meanchey. It was beyond the parameters of this research to objectively assess the effectiveness of forest protection efforts, but interviews indicate that threats to the forest have significantly diminished since the forest has been under the monk's protection, and that only a small portion of the MCF, to the south of the Ochea River, is still subject to some illegal logging and hunting during the dry season.

The MCF team believes their success is due to consistent patrolling, combined with their outreach and awareness efforts with local communities and with provincial and local government departments. In addition, as of July 2008, the military presence near the MCF was reduced considerably. Logging by the army dramatically decreased when, just prior to the country's parliamentary elections, Prime Minister Hun Sen issued orders to provincial authorities to crack down on illegal logging. In addition, all spare soldiers were sent to the Thai-Cambodian border to strengthen Cambodia's position in the conflict that began in October 2008 with Thailand over the ownership of Preah Vihear Temple. The military area located near the southwest boundary of the MCF was also moved elsewhere.

In 2007, CFI, in conjunction with Terra Global Capital (TGC), began designing the first carbon offset project under the United Nations' strategy for Reducing Emissions from Deforestation and Forest Degradation (REDD) for community forests in Oddar Meanchey. Due to CFI's effort and FA cooperation, in November 2008, the Ministry of Agriculture, Forestry, and Fisheries (MAFF) recognized the boundaries of twelve potential community forests in Oddar Meanchey. It then issued a ministerial regulation in May 2009 that legally designated thirteen community forests, including the MCF.

Community forests fall under the authority of the FA/MAFF. They are designed to enable forest-dependent communities to regenerate degraded forests or protect existing forests, and to manage them sustainably to support their livelihoods. This is particularly important in a province like Oddar Meanchey, where commercial logging has been so destructive and where valuable land continues to be taken by the elite, and lost to the majority of citizens who depend on natural resources for their survival. Under community forest guidelines, the communities establish management committees made up of elected members from their villages, and carry out patrols of the forest to monitor illegal logging, land clearing, and hunting. They have access to the forest for resources such as resin, bamboo, and fish, which they can harvest for subsistence use or for selling. The MCF follows these guidelines, but it was the monks, not the villagers, who spearheaded the community forest project.

In 2011, the project received validation from the Voluntary Carbon Standard (VCS) to sell its carbon on the international market. The project is being implemented by the Cambodian Forestry Administration, Pact Cambodia, and the Children's Development Association, a local NGO. It is hoped that this project will provide much of the necessary funding to continue the conservation and forest management work for these communities. Because of the area's new legal community forestry status, and the prospective international carbon offset financing, there is now considerable pressure on government officials, military, and other stakeholders to respect the boundaries of the MCF and to contribute to its protection.

In chapter 3, figure 1 shows the thirteen community forests of Oddar Meanchey Province, of which the MCF is the largest (another CF has since been added, for a total of fourteen). The MCF is found approximately forty kilometers southeast from the provincial capital of Samraong Town, and thirty kilometers from Anlong Veng, in Koun Kriel Commune in Samraong District. An additional six-kilometer "wildlife corridor" runs from the northern boundary of the MCF to the Thai border.[16]

LESSONS LEARNED

This section considers why the Samraong monks have been able to achieve what appears to be real and lasting conservation outcomes, with far fewer resources than government agencies and local NGOs and, arguably, with a more sustainable impact. It also considers the great potential for conservationists working towards the protection of Cambodia's natural resources to support, if not directly join forces with, newly engaged Buddhist movements such as the MCF in ways that can be mutually beneficial. This discussion hopes to highlight some of the issues to consider in this type of partnership, calling attention to the potential benefits and difficulties that can arise.

BENEFITS OF PARTNERSHIP

Supporting Buddhist-led conservation initiatives can complement current conservation efforts while encouraging Khmer leadership and civil society responses to natural resource depletion. Currently, there are only a handful of engaged Buddhist projects associated with conservation in Cambodia. Most monk-led projects focus on development and human rights. The MCF is the only example in the country of monks actively patrolling and managing a protected forest area. Elsewhere, Buddhist monks are involved in environmental education through pagodas and other forms of community forestry.

Ven. Sopheap (now disrobed) runs the country's only Cambodian Buddhist conservation NGO, the Association of Buddhists for the Environment (ABE). He and his team have been working since 2005 through pagodas and networks of monks in various provinces to raise environmental awareness and to promote the conservation of Cambodia's natural resources. ABE was originally supported by two NGOs, Alliance of Religions and Conservation (through a joint World Bank project) and Mlup Baitong, and is primarily focused on environmental awareness. It has had considerable success in running environmental education projects in several provinces in Cambodia, and in developing a network of monks and pagodas involved in conservation. Their activities include training monks and villagers in tree planting, vegetable gardening, and the creation of fishponds around pagodas; tree ordination ceremonies; and developing village microfinance schemes, discussion forums and films, and environmental youth groups.

Ven. Nem Kim Teng heads another effort, the Sante Sena community forestry project based out of Wat Prey Chlak, in Svay Rieng Province, and he has started a network of monks involved in community forestry around the country. Another Buddhist initiative is the Dhammayietra peace walks, the first of which was led in 1992 by the Venerable Maha Goshananda, a revered figure known as "Cambodia's Ghandi." The peace marches call attention to a variety of social issues including deforestation, and are one of the first examples of an indigenous movement linking Buddhism and environmentalism. Since Ven. Goshananda's death in 2007, one of his disciples, a woman named Oddom Van Syvorn, has been leading the walks. This "Lady Gandhi" organizes a walk every year in a different province, and focuses on several themes, including respecting the Five Precepts; building awareness of HIV/AIDS; and spreading an ethic for environmental protection, specifically about the need to plant trees and respect nature. There are likely several other small-scale projects run by Buddhists focused on the environment that are not covered here. What they all have in common is that they are led by charismatic individuals. These leaders should be nurtured as much as possible to strengthen their activities and to inspire others to follow suit.

Although the number of initiatives is still small, the potential for replication and expansion of Buddhist-led conservation activities is significant in what is almost entirely a Buddhist country. The successful MCF project is one model that can be expanded and replicated in Oddar Meanchey Province and beyond.

As previously noted, the MCF is one of fourteen community forests in Oddar Meanchey Province. Together, these forests make up one of Cambodia's first REDD project sites, which cover over 68,000 hectares of community forest, and with over 20,000 people living within and around these boundaries. The MCF is the only community forest (CF) within the REDD project site where monks are participating in protection. Members of the other CF committees, having witnessed the success of the MCF, are now seeking advice from the monks on how best to protect their forests, and have expressed the desire to have monks join their protection efforts. These CFs are under much greater threat from deforestation and land appropriation than the MCF. The opportunity is ripe to expand the model of Buddhism and conservation within the entire CF pilot site: Some fifty monks live in seven different pagodas in the greater CF area, many of whom are keen to participate in forest protection efforts and to receive training from the MCF team.

CHALLENGES

In addition to the potential benefits that can arise from partnerships between conservation groups and engaged Buddhist groups, there are also various potential pitfalls and difficulties that could hamper these relationships. It is important to consider some of these.

> There is a lot of criticism. People here say the monks should stay in the temple, learn Dharma, go on alms collections, focus on meditation and attaining nirvana, and don't get involved in problems. This is because they don't understand the role of monks. We changed that. Monks have to support local people.[17]

The words of this monk illustrate the conflict that exists in Cambodia between engaged Buddhism and its detractors. The monk continues:

> Once they see what we are doing, when they see us becoming involved in problems within the community and trying to help, they stop criticizing and bring us offerings of food. So we have changed their concept of what a monk is, and show them that a monk should support local people.

Many of the Samraong monks echoed this sentiment. They feel that the people who criticize them for their work do not really understand the role of monks in society, and do not understand Buddhism. Their critics are often those who stand to lose from forest protection because they want to appropriate land or cut timber to sell commercially. Above all, the monks emphasized that their involvement with forest protection was not against the Buddhist monastic rules. In the words of the provincial head monk, "We feel we are following Buddha's advice."

Despite some apprehension about having monks tasked with management and administration of protection efforts, most people interviewed felt the monks should remain involved in patrols and operations in order to safeguard the forest. This research did, however, bring up questions related to the balance between community, monks, conservation groups, and authorities in any conservation partnership.

TRADITIONALISTS VERSUS MODERNISTS

In addition to the concerns expressed by some local people about the monks' environmental activities, it is also important for conservation groups to be aware of some of the historical tensions between engaged Buddhists and the more traditional elements of the religious and political establishment in Cambodia.

This tension has been between the "traditionalist" and "modernist" strains of the Cambodian *sangha*, which itself is divided into two branches, the majority Mahanikay, and the minority Thommayut, the latter introduced from Thailand in the ninteenth century and still associated with the royal family.[18] Both branches have traditional and modernist elements, but engaged Buddhism tends to be associated with the modernist Mahanikay tradition (as are the monks of the Samraong Pagoda).

Cambodia's ruling party, the Cambodian People's Party (CPP), is for the most part associated with traditional Mahanikay Buddhism, and is wary of modernist monks. It encourages a pagoda-based role for monks that isolates them from secular affairs and focuses instead on helping the Cambodian people live in accordance with Buddhist moral principles.[19] The state can feel threatened by political participation of monks because of their influence over the general populace, exemplified by their pivotal role in the anticolonial protests of the 1940s and 1950s that helped topple the French government.[20] The ruling party's suspicion of modernist Buddhism can be explained further by its connection to non-CPP, non-Cambodian influences. The king's support of engaged Buddhist projects perhaps only further antagonizes the Cambodian government, given the tension between the royal family and the CPP.[21] Another reason for official discomfort with the intertwining of religious and secular issues is the attention it calls to some of the more controversial aspects of Cambodian governance, such as deforestation, weak land rights for local people, and lack of government action concerning the rise of HIV and human trafficking, to name a few.[22]

Some Cambodians view engaged Buddhism as a Western, or at least non-Cambodian, phenomenon. This is partly the result of financial and resource support for Buddhist social development work from foreigners, international NGOs, and foreign governments. The scholar Ian Harris cautions, "To put the matter simply, external financial support is given exclusively to forms of Buddhism moving in a socially engaged, modernist direction." Given the Cambodian government's wariness of foreign, especially Western, influence over the country's national affairs, Harris cautions that the Mahanikay modernists "must strive to avoid the charge that they have become nothing more than servants of the international community."[23] He sums up the tension arising from foreign support of engaged Buddhism, and how it threatens to widen the gap between modernist and traditionalist strains within the country:

> From what we know about the history of Buddhism in Cambodia, it is only a matter of time before the traditionalists will react to what they regard as a foreign-financed and potentially unpatriotic segment within the monastic order, and this could be quite damaging to the immediate prospects of a Buddhist revival in the country.[24]

In summary, there are many benefits to partnerships between international NGOs, local NGOs, and engaged Buddhist groups, but it is also wise to proceed cautiously in light of some of the issues raised here. Despite some apprehension by traditionalists about Cambodia's new engaged Buddhist movements, it seems likely that Buddhist involvement in social development activities is a trend that will only increase, as it has across the Asian region.

CONCLUSION

This chapter has highlighted a model of engaged Buddhist action designed to protect threatened forests, one that can be replicated and expanded throughout Cambodia. With few resources, the monks of Samraong

Pagoda have proven themselves to be powerful conservationists: they have acquired legal protection of an 18,261-hectare forest, established effective forest patrols, demarcated forest boundaries, raised environmental awareness among local communities, organized community volunteers, linked with government authorities and NGOs, and by all accounts, significantly reduced forest crime in the MCF. In addition, they have attracted foreign support and are now part of the first REDD project in Cambodia that promises to provide sustainable funding for management and protection of the MCF over the long term.

It is important to note that these impressive achievements sprang from the belief that by protecting the MCF, the monks are following the Buddha's example and the principles he set out in his teachings. They believe that their forest protection activities are perfectly aligned with monastic aspirations to eliminate the suffering of all beings and to live ethically by obeying the moral precepts. Living a moral life according to virtues such as nonviolence, compassion, generosity, and simple living, and reaping the good merit this will incur, has been the main motivational force behind the monks' forest protection activities. They also point to the close connection of the Buddha's life with nature, and how the major events of his life—birth, enlightenment, and death—all occurred in the forest.[25] The monks also protect the forest to safeguard the abodes of spirits, and to maintain forests for future generations, which they see as their patriotic duty to Cambodia.

Buddhist belief has also been a powerful motivator for other key stakeholders, particularly the Buddhist communities living near the MCF. The monks have been able to spark and cultivate a conservation ethic among these villagers by explicitly linking nature conservation to the life of the Buddha, and to Buddhist principles such as karma and moral living.

The MCF example has demonstrated that Buddhist monks can be powerful messengers for environmental protection. This is particularly so in a country where most authority figures have lost legitimacy due to corruption and policies that marginalize the majority of the Cambodian population. Monks have always acted as moral and spiritual guides in their communities, and their moral authority is powerful and respected. When monks extend their efforts to forest protection, it lends forest conservation a legitimacy that is often weak in foreign-supported conservation projects. At the same time, the MCF's activities could be greatly enhanced by external financial and technical support. The example of the MCF shows that monks can be strong conservation allies, and that there are real opportunities for collaboration in ways that could be mutually beneficial, if some of the potential difficulties that can arise are mitigated.

The monks have developed some unique approaches to law enforcement based on Buddhist principles that can be adopted by other conservation projects in Cambodia. They have developed a system of "soft" enforcement in which violence is minimized. This approach involves unarmed patrols, treating offenders without anger, a "three-strikes-you're-out" policy, the setting aside of an ordained forest, and the use of photography and large groups of monks on patrol to deter offenders. This is complemented by strong relationships with local, provincial, and national authorities; with NGOs; and with local villagers. The MCF's protection system has demonstrated the effectiveness of including monks in monitoring and patrolling the forest, and their impact on deterring forest crime.

Buddhists across Asia are mobilizing in all kinds of ways to protect the environment, and they are using Buddhist principles as fuel for their conservation actions. This trend is just dawning in Cambodia and holds great potential for spreading conservation messages and enhancing on-the-ground protection. Additional research on engaged Buddhist movements and their conservation activities would be of great value to this growing movement.

NOTES

1. Dennis McMahon, "Assessment of Community Forestry Sites and Migration Patterns in the Oddar Mean Chey Province, Cambodia" (Study Conducted for Community Forestry International, Phnom Penh, March 16, 2008), 10.

2. Ibid., 21.

3. "Factsheet: Community Forestry Carbon Offset Project" (Phnom Penh: Community Forestry International, June 2008).

4. Amanda Bradley, "Communities and Carbon: Establishing a Community Forestry-REDD Project in Cambodia" (Phnom Penh: Pact, 2009), 5.

5. Amanda Bradley and Robert Oberndorf, "Buddhism and the Role of the Pagoda in Community Forestry Development in Cambodia" (Report by Community Forestry International, Phnom Penh, March 2005), 3.

6. Robert Walker, "Cambodia: A Land Up for Sale?" (BBC online article, on <http://news.bbc.co.uk/1/hi/world/asia-pacific/8144130.stm> [accessed May 18, 2009]).

7. "Cry From The Forest: A 'Buddhism and Ecology' Community Learning Tool" (Phnom Penh: The Buddhist Institute in cooperation with the NGO Working Group for Non-formal Monk Environmental Education Project [MEEP] and with support of UNDP-ETAP in collaboration with UNESCO, 1999), 30. Tree ordination in Cambodia was borrowed from the Thai practice.

8. US State Department information on Cambodia online on <http://www.state.gov/r/pa/ei/bgn/2732.htm> [accessed June 3, 2009].

9. Damien Keown, *Buddhist Ethics: A Very Short Introduction* (Oxford: Oxford University Press, 2005), 14.

10. Buddhist monk interview.

11. Judy Ledgerwood, "Buddhist Practice in Rural Kandal Province 1960 and 2003," in *People of Virtue: Reconfiguring Religion, Power and Moral Order in Cambodia Today*, ed. Alexandra Kent and David Chandler, Studies in Asian Topics, no. 43 (Copenhagen: NIAS Press, 2008), 155.

12. Ibid., 155.

13. Keown, *Buddhist Ethics*, 39.

14. Pragati Sahni, *Environmental Ethics in Buddhism: A Virtues Approach* (New York: Routledge, 2008), 141.

15. Alexandra Kent and David Chandler, eds., *People of Virtue: Reconfiguring Religion, Power and Moral Order in Cambodia Today,* Studies in Asian Topics, no. 43 (Copenhagen: NIAS Press, 2008), 310.

16. "Factsheet: Community Forestry Carbon Offset Project" (Phnom Penh: Community Forestry International, June 2008).

17. Buddhist monk interview.

18. Ian Harris, *Cambodian Buddhism: History and Practice* (Bangkok: University of Hawaii Press, 2005), 213.

19. Ibid., 242.

20. Ibid., 245–49.

21. Ibid., 212.

22. Ibid., 209.

23. Ibid., 213.

24. Ibid., 212.

25. Donald Swearer, "Buddhism and Ecology: Challenge and Promise," Harvard University, <http://fore.research.yale.edu/religion/buddhism/index.html> [accessed 25 July 2009].

THE ETHNOECOLOGY OF THE KAVET PEOPLES IN NORTHEAST CAMBODIA

IAN G. BAIRD

RESEARCHERS, PROTECTED AREA MANAGERS, AND DEVELOPMENT workers are frequently interested in the ways that local people use and manage different species and ecosystems. But much less is known about how different peoples conceive of and classify habitats or ecosystems in their own languages. This is especially the case in Southeast Asia, where despite a very wide range of peoples and languages, very little research of this nature has been done. Yet these classification systems are the spatial foundation for human interactions with nature, they are the basis for livelihoods, and they are integral parts of peoples' cultural heritage. When people live inside or near protected areas (PAs), understanding how they conceptualize and classify nature is crucial, as this knowledge can help indicate ecological relationships, and assists in better understanding the ways that people interact with nature. Thus, this knowledge can act as a tool for conservation and sustainable resource use.

This chapter illustrates the importance of ecological classification systems at the scale of habitat types in relation to natural resource management—and, more specifically, PA management—by reporting on the findings of ethnoecological research conducted with the Kavet subgroup of the Brao ethnic group in Kok Lak Commune, Veun Sai District, Ratanakiri Province, northeast Cambodia.[1] The practical value of understanding indigenous ecological classification systems is discussed, including their usefulness in local land use planning and PA management. The swidden agricultural system of the Kavet in Kok Lak is considered in the context of the management of Virachey National Park (VNP). Some Kavet from Kok Lak make a strong argument that they should be allowed to conduct swidden agriculture inside VNP and in low mountain areas adjacent to the PA, a proposition that is considered here.

THE BRAO AREA

The Brao[2] are a Mon Khmer Western Bahnaric language–speaking group[3] with a long history of inhabiting the remote, hilly, and densely forested region that straddles Attapeu Province in southern Laos and adjacent parts of Stung Treng and Ratanakiri Provinces in northeastern Cambodia. Their territory composes almost all of VNP.[4] The total worldwide Brao population is approximately 60,000, with more than half residing in northeastern Cambodia and most of the rest living in southern Laos. There is also one Brao village in the Central Highlands of Vietnam.[5]

In Cambodia, the Kavet, one of the nine well-recognized Brao subgroups, inhabit Kok Lak Commune, Veun Sai District, Ratanakiri Province, and part of Siam Pang District, Stung Treng Province.[6] The Kavet make up approximately 2.7 percent of the population of Ratanakiri.[7] Although they are generally considered to be ethnically distinct from the Umba Brao subgroup who inhabit Taveng District in Ratanakiri Province, the Kavet speak a mutually intelligible dialect of the northern branch of the Brao language.[8]

Kok Lak Commune is situated north of the Sesan River. Bordering Laos's Attapeu Province to the north, the largely forested and relatively remote commune is almost completely populated by Kavet peoples living in the villages of Heulay (Lalay), Rok, 'Ntrak (Drak), and La Meuay.[9] All of the northern and much of central Kok Lak are included within VNP, extending up the border with Laos.

In November 1993, the Cambodian government designated twenty-three national PAs covering 3,327,200 hectares into four groups. These included seven national parks, ten wildlife sanctuaries, three protected landscapes (including forests around Angkor), and three multiple-use areas.[10] VNP is Cambodia's largest national park and one of its most expansive PAs, with an area of 332,500 hectares in Siam Pang District, Stung Treng Province, and in Veun Sai, Taveng, and Andong Meas Districts, Ratanakiri Province.[11] VNP is bordered to the north by Laos's Attapeu Province and in the east by Vietnam's Kon Tum and Gia Lai Provinces, and is surrounded on two sides by major rivers, the Sekong to the west and the Sesan in the south. The elevation of the park ranges from approximately 100 meters above sea level near the Sesan and Sekong Rivers to over 1,000 meters above sea level on the high mountain ranges that extend along the Laos-Cambodia border.[12]

VNP supports populations of a number of regionally and globally threatened and endangered wildlife species, and the park is considered to be important for biodiversity conservation.[13] The predominance of dense forest is apparently made possible by the over 2,000 millimeters of precipitation that falls in most of the park.[14] Within the core zone and in the northeast, there are some "abandoned shrublands" (i.e., former swidden plots).[15] Although there are presently no villages situated inside VNP,[16] due to displacement during the Khmer Rouge period and, most recently, because of pressures by park officials,[17] indigenous Brao (Umba, Kavet, and Kreung subgroups) peoples historically lived throughout much of the PA, and consider most of VNP to be their homeland.[18]

Historically, all Brao village communities had control over territories in which they moved their villages, and especially their family-owned swidden fields. They often maintained boundaries between the village territories of other communities in order to avoid spatial taboos that prohibited the swidden fields from one community from crossing over the swidden fields of another.[19] Many ethnic Brao villages are inside the park's unofficial buffer zone, and their inhabitants make up the majority of the over 12,000 people living in nine communes and forty-one villages located near the park.[20] People living around VNP continue to hunt, fish, gather forest products, and engage in swidden agriculture within the PA.[21] The boundaries of VNP were set using topographical maps and aerial photographs in Phnom Penh, and there was little consideration of the human ecology of the area at the time that the park's boundaries were established.[22]

Figure 1. Map of Approximate Locations of Brao Subgroups in Northeast Cambodia and Southeast Laos

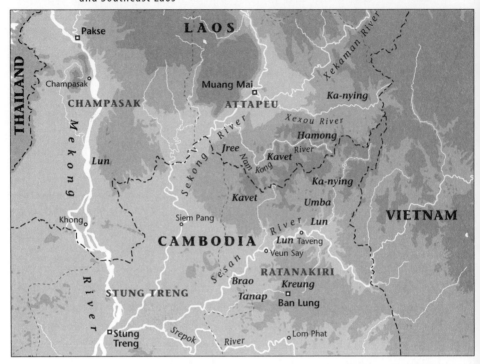

Most of the Kavet who previously lived inside VNP were forcibly evicted by successive Cambodian governments, beginning with Norodom Sihanouk's *Sangkum Reastr Niyum* (People's Socialist Community) government in the early 1960s, the Khmer Rouge in the 1970s, the Vietnamese-backed Hun Sen government in the early to mid-1980s, and finally by park staff receiving financial support from the World Bank in the early 2000s.[23] Most were relocated to lowland areas near the Sesan and Sekong Rivers. Despite these various attempts to make the Kavet into lowlanders, many have intermittently—and for various reasons—tried to move back into the uplands.[24] The four villages in Kok Lak Commune are presently situated away from the Sesan River next to the Lalay Stream, although they are still in the lowlands.

The factors presently affecting the management of the part of VNP located in Kok Lak Commune are complex, involving a variety of intertwined issues associated with PA management, community development, land use and natural resource management, logging concessions, agriculture, politics, culture, and security.[25] For many years, many Kavet have been lobbying to be allowed to return to their original homelands inside VNP to conduct swidden agriculture, but have so far been discouraged by the government due to concerns related to security, the provision of government services to villagers, and, most recently, PA management.[26]

The role of Kavet agriculture, hunting, fishing, and gathering of forest products in the park and nearby areas has been of great interest to outside observers and government officials,[27] largely because the Kavet are considered to be a key group when it comes to PA management. Moreover, the Kavet have, so far, generally adapted poorly to having been removed from their traditional lands inside the park, as there is insufficient paddy farmland for them in the lowlands, and many are still uncomfortable with lowland farming methods. They suffer from chronic food shortages, and the commune is considered to be one of the poorest in Cambodia. Many Kavet feel that their own lands are in the uplands, not in the lowlands.

ETHNOECOLOGY

Ethnoecology has different meanings to researchers. F.L. Dunn defined it as "how people evaluate, classify, label, and reject or use all the resources, biotic and nonbiotic, of their ecosystem."[28] D.A.Posey considered ethnoecology to be "indigenous perceptions of 'natural' divisions in the biological world and plant-animal-human relationships within each division."[29] Posey contended that these cognitively defined ecological categories do not exist in isolation; thus, ethnoecology must also deal with the perceptions of interrelatedness between divisions in nature.[30] More recently, N. Barrera-Bassols and V.M. Toledo have broadly defined ethnoecology as "an integrative study of beliefs, knowledge, and practice of a given social entity."[31]

An "ecosystem" is "a biological community together with its associated physical environment,"[32] or "the sum total of vegetation, animals, and physical environment in whatever size segment of the world is chosen to study."[33] Ecological zoning differs from place to place and for different people, but space, time, and hierarchical levels, as well as livelihoods, fundamentally influence ecological perceptions and classification systems.[34] It remains unclear how much the way we classify nature is based on "universal principles" inherent in all peoples, rather than on cultural and linguistic influences and life experiences.[35] Yet the question is important when considering natural resource management planning.[36] If indigenous peoples, such as the Kavet, see ecosystems in fundamentally different ways than PA managers, confusion and associated management problems are apt to arise. This may partially explain why indigenous peoples and PA managers frequently come into conflict when it comes to park zoning.[37]

Invariably there are important reasons why indigenous peoples divide ecological areas in particular ways.[38] Even a partial understanding of local ecosystem classification systems can provide PA managers with crucial clues that can benefit natural resource management planning and implementation.[39] Rural inhabitants are often the owners of the most complete knowledge bases about the unique part of the world where they have lived for generations.[40]

ECOLOGICAL CLASSIFICATION SYSTEMS

There are essentially six domains that influence the ways in which ecological zones are classified:

1. Landforms and geographical localities

2. Soils

3. Climate

4. Vegetation types

5. Ecological succession

6. Human land and resource use[41]

A landscape can be described as a heterogeneous land area composed of an interacting mosaic of habitats, ecosystems, and land uses, and many researchers believe that landscapes—sometimes referred to as ecoregions—are crucial to maintaining biodiversity.[42] A variety of associated biotopes, or microhabitats, are combined to make up an ecoregion, or broad landscape.[43]

The role of "culturally influenced" ecological succession is crucial to forest management,[44] and it is increasingly recognized that the view that forests are pristine wildernesses is largely socially constructed. Instead, most forests have been greatly influenced by human practices over generations. Thus, the way the Cambodian Kavet view and manipulate successional patterns is important to understand, and is certainly related to their ecological classification systems. Analyzing temporal and spatial influences on ecological areas and associated management strategies is generally the key to understanding ecological classification systems.[45]

METHODOLOGY

The research began in July 1999, with a two-day "Participatory Landscape Ecology, Wildlife Habitat Assessment, and Mapping Training Workshop," which was organized in Veun Sai District. Officials from the Ministry of Environment and VNP attended, along with approximately twenty wildlife and forest "experts" selected by their peers from Kok Lak's four communities. The workshop was designed to introduce research objectives and methodologies to officials and villagers, and to initiate PA staff in the collection of ethnoecological information using participatory methods. The workshop represented the beginning of the collection of ethnoecological data in cooperation with the Kavet.

Most of the methods introduced were adapted from participatory rural appraisal (PRA) techniques. These included "free listing" activities to generate an indigenous list of Kavet-language "forest" types and "ranking" exercises designed to investigate forest habitat and agriculture suitability associations.[46] "Habitat/season wildlife usage matrices" were also introduced to provide information about seasonal wildlife behavior,[47] as well as "PRA mapping" to investigate spatial relations and resource use patterns.[48]

Following the initial workshop, a team of researchers that included the author and a number of Kavet counterparts conducted a seven-day field survey in Kok Lak's four villages. Semistructured interviews and open discussions were initiated with groups of local experts (mainly middle-aged men, as well as some women) in order to determine how many ecological areas, or biotopes, are generally recognized by the Kavet, and what characteristics are used to classify ecological areas. Finally, in

February 2000, field surveys in Kok Lak Commune were conducted in two stages.

Other researchers have stressed the importance of collecting ecological information from indigenous peoples during forest walks, and have emphasized the value of spending as much time as possible in the forest with native peoples.[49] I concur. When in the forest, the confidence of the Kavet increased significantly, enhancing their ability to articulate ecological concepts and processes. Our forest treks proved to be an invaluable part of the research, and greatly helped increase understandings of Kavet land use patterns and ecological classification systems. During walks in the forest, we generally trekked until we reached habitat types that appeared different from areas already passed. We asked male Kavet consultants with us what they called the area in their own language.

A rapid visual habitat assessment was conducted, adapted from the system developed by R. Steinmetz.[50] These rapid assessments were based on easily recognizable characteristics in ecological structure (see table 1 for a list of the categories assessed and scoring ranges). We discussed the successional patterns and history of human use with the Kavet, as well as area significance. The Kavet also identified dominant and ecological keystone plants, and provided information about Kavet uses of particular species, the suitability of the area for swidden agriculture, and other things relevant to understanding land use patterns and management practices. Wildlife usage and abundance were discussed, although we did not focus on this. Instead, understanding the biotope classification system of the Kavet was key. Each Kavet habitat type was photographed.

Table 1. Habitat Parameters used in Kok Lak Fieldwork

#	Habitat Parameter	Scoring Range
1	Canopy cover	1 to 4 (open crown to overlapping)
2	Canopy height	In meters
3	Ground moisture	1 to 4 (dry to moist)
4	Ground cover	1 to 5 (very low to ground covered)
5	Woody climbers/vines	1 to 4 (rare to abundant)
6	Slope of land	1 to 5 (flat to steep)
7	Visibility	1 to 5 (open to thick)

8	History of disturbance	Descriptive
9	Soils	Descriptive
10	Altitude at sea level	In meters
11	Canopy layers	Number of layers
12	Dominant species	Kavet name of selected species
13	Mixture of species	Descriptive
14	Notes	Descriptive

Microhabitat information is important not only for understanding habitat diversity in VNP, but also for getting a feel for how the Kavet classify the forest and make land use decisions. Clearly all this information is crucial for PA management.

THE KAVET ECOLOGICAL CLASSIFICATION SYSTEM

Kavet ecological classification systems are essentially based on two general ecological terms: *bree* and *dak*. *Bree* is often translated as "forest" in English, or *prey cheu* in Khmer. However, Gerard Diffloth, a leading Mon-Khmer-language linguist, believes that *bree* does not actually mean "forest," but "the condition of the land." A *bree* can be a forest area, since forests often dominate "the condition of the land," but the word *bree* can also describe areas that are not normally categorized as "forests" by foreigners or lowland Khmers. For example, the Kavet call grasslands *bree treng* (short vowel) and salt licks are known as *bree graik*. Areas covered entirely in flat slabs of granite, with virtually no vegetation present, are known as *bree ta-tar*, or *bree ta maw ta tar*. *Bree* types are sometimes named after dominant species of plants, but *bree* is not used to describe individual plant species. Instead, *bree* types represent ecological areas, known to ecologists as biotopes.

Dak localities are limited to rivers and streambeds up to the tops of their banks. *Bree* terms exist for all terrestrial lands, and *dak* areas make up all the aquatic habitats. *Dak* means "water" or "the condition of the water." Therefore, the Kavet do not classify seasonally inundated forests found in a riverbed as a type of *bree*, because even though they are "forests," they are found in essentially *dak* areas. However, these forests found in streambeds are not named using the word *dak*. Instead, the association with water seems to be implied and unspoken. Seasonally inundated forests in river-beds are therefore called by the names of trees. However, areas adjacent to but outside of river and streambeds that are flooded annually are considered *bree* areas, regardless of the extent of flooding annually. These areas include *bree trang* (pandus grass forests flooded annually) and *bree a-ra cheu* (a type of bamboo forest found in lowland floodplains). *Bree ja-naw*, or stream forest, is found only outside but adjacent to streambeds, on the top of stream banks.

"BIOTOPES"

During initial consultations with the Kavet, 106 *bree* habitat types were documented in Kavet language, far more than expected. However, some new classes were later added, others were revised, and some were removed from the list after further investigation. Table 2 lists 108 Kavet basic ecological classification categories, or biotopes, of *bree*. There are twenty-one *bree* types considered to represent "broad ecological classes," twelve based on "topographical" or landform features, six representing "successional" stages, seven founded on "pedologic" or soil characteristics, fifty-seven linked to "dominant plant species" and five "miscellaneous" types (see table 2).

Table 2. Base Kavet Ecological Classification Terms

No	Kavet Name	English Name	Class Type
1	Bree lawng	Semi-evergreen	Broad class
2	Bree lawng mee mee	Pure tree forest	Broad class
3	Bree lawng hundeum hooan	Tree seedling forest	Broad class
4	Bree lawng hundeum geut	Large tree seedling forest	Broad class
5	Bree lawng kreung	Old semi-evergreen forest	Broad class
6	Bree gra	Old semi-evergreen forest	Broad class

7	*Bree kreung*	Old semi-evergreen forest	Broad class
8	*Bree gloong*	Wide area of primary forest	Broad class
9	*Bree jyeul*	Spring or wetland forest	Broad class
10	*Bree lip-lip*	Thick vine-dominated forest	Broad class
11	*Bree lip grit*	Very thick vine forest	Broad class
12	*Bree breuit*	Very thick vine forest	Broad class
13	Bree bok doon	Thick vine and grass forest	Broad class
14	Bree chong charong	Open forest	Broad class
15	Bree reung-reung	Deciduous very open forest	Broad class
16	Bree treng (short vowel)	Open grassland field	Broad class
17	Bree treng mee mee (long vowel)	Dipterocarp forest	Broad class
18	Bree baw-jun-trooay-lawng	Mixed bamboo/tree forest	Broad class
19	Bree lawk-gra-law	Island forest	Broad class
20	Bree glim	Cold forest	Broad class
21	Bree do	Hot forest	Broad class
22	Bree jundoo	Mount. semi-evergreen forest	Topographic
23	Bree jyeung jundoo	Foot of mountain forest	Topographic
24	Bree banoi-gawl	Long mountain range forest	Topographic
25	Bree dol-dol	Small hill forest	Topographic
26	Bree wang-hoong	Open forest at bottom of forest	Topographic
27	Bree loo-loo	Depression area forest	Topographic
28	Bree bawl ja-naw	Stream edge forest	Topographic
29	Bree koot-ja-naw	End of stream area	Topographic
30	Bree treup-treup	Flat land forest	Topographic
31	Bree ga-leung	Slight sloped forest	Topographic
32	Bree chang	Moderately sloped forest	Topographic
33	Bree ja-rung-ja-rung	Steep sloped forest	Topographic

34	Bree char-char	2–4 year fallow swiddens	Successional
35	Bree char gra	5–10 year fallow swiddens	Successional
36	Bree char-chawng	11–70 year fallow swiddens	Successional
37	Bree gar-yawng	Flowering bamboo forest	Successional
38	Bree heu-toong	Young bamboo forest	Successional
39	Bree baw-brum-brum	Dead bamboo forest area	Successional
40	Bree ta-maw	Moderate-size rock area	Pedologic
41	Bree gao	Huge boulder area	Pedologic
42	Bree tar-tar	Rock slab area	Pedologic
43	Bree phaik-phaik	Sandy soil forest	Pedologic
44	Bree pre-bree-o-bree-o	Pebble soil forest	Pedologic
45	Bree jundoo pa-door	Bluish crystal rock forest	Pedologic
46	Bree ga-wang	Large boulder forest	Pedologic
47	Bree jya	Good soil for swidden forest	Miscellaneous
48	Bree rang-ha-rang	Poor quality forest for agr.	Miscellaneous
49	Bree mee-mee	Poor soil for swidden forest	Miscellaneous
50	Bree grait-grait	Salt-lick	Miscellaneous
51	Bree dom	Old semi-evergreen forest	Miscellaneous
52	Bree ara-cheu	Bamboo forest	Dominant sp.
53	Bree eung-le	Bamboo forest	Dominant sp.
54	Bree lao	Bamboo forest	Dominant sp.
55	Bree baw	Bamboo forest	Dominant sp.
56	Bree ha-tee-ang-mat	Bamboo forest	Dominant sp.
57	Bree ha-tee-ang-ja-reum	Bamboo forest	Dominant sp.
58	Bree pok	Bamboo forest	Dominant sp.
59	Bree ha-goo	Bamboo forest	Dominant sp.
60	Bree taw-nawp	Bamboo forest	Dominant sp.
61	Bree gooan-bawk	Bamboo forest	Dominant sp.
62	Bree chao	Bamboo forest	Dominant sp.

63	Bree ha-ling	Bamboo forest	Dominant sp.
64	Bree baw-ra	Bamboo forest	Dominant sp.
65	Bree baw-ja-reum	Bamboo forest	Dominant sp.
66	Bree lao-lan	Bamboo forest	Dominant sp.
67	Bree tra-lao	Bamboo forest	Dominant sp.
68	Bree ha-doong	Vine forest	Dominant sp.
69	Bree trang-trang	Pandus grass forest	Dominant sp.
70	Bree arawng	Pandus grass forest	Dominant sp.
71	Bree ta-ngaik	Imperata grassland	Dominant sp.
72	Bree kreung-ga-bawng	Dipterocarp forest	Dominant sp.
73	Bree ha-buk	Rattan forest	Dominant sp.
74	Bree ja-weuk	Rattan forest	Dominant sp.
75	Bree jawn-jaw	Rattan forest	Dominant sp.
76	Bree ja-woo-at	Rattan forest	Dominant sp.
77	Bree a-rae	Rattan forest	Dominant sp.
78	Bree tam-pawng	Rattan forest	Dominant sp.
79	Bree an-ton	Rattan forest	Dominant sp.
80	Bree gooan-eel		Dominant sp.
81	Bree ga-lawk	Palm forest	Dominant sp.
82	Bree ja-lawk	Palm forest	Dominant sp.
83	Bree beek	Palm forest	Dominant sp.
84	Bree an-gee-ak		Dominant sp.
85	Bree ja-roo	Palm forest	Dominant sp.
86	Bree ja-nap		Dominant sp.
87	Bree ga-naing	Fern-like vine forest	Dominant sp.
88	Bree gray-gray	Fern-like vine forest	Dominant sp.
89	Bree ya-yoot	Fern forest	Dominant sp.
90	Bree an-to-eng	Fern forest	Dominant sp.
91	Bree an-heng		Dominant sp.
92	Bree roon	Wetland forest	Dominant sp.

93	*Bree ta-jak*	Wetland reed forest	Dominant sp.
94	*Bree ta-cheu-ay*	Rattan wetland forest	Dominant sp.
95	*Bree nee-ang-ga-nee-ang*	Wetland forest	Dominant sp.
96	*Bree jyawk*		Dominant sp.
97	*Bree jara*		Dominant sp.
98	*Bree treuay*	Vine forest	Dominant sp.
99	*Bree ja-meu-lat*	Thick understory vine forest	Dominant sp.
100	*Bree ja-meu-ga-trawk*	Vine forest	Dominant sp.
101	*Bree graw-mat*	Wild banana forest	Dominant sp.
102	*Bree anawng*	Semi-evergreen forest	Dominant sp.
103	*Bree chawk-cha-nawl*	Pine forest	Dominant sp.
104	*Bree lawng-ta-bak*	Deciduous open forest	Dominant sp.
105	*Bree graw-daw*		Dominant sp.
106	*Bree preung-teeap*		Dominant sp.
107	*Bree gun-daw*	Climber forest	Dominant sp.
108	*Bree ha-jawng*	Malva nut forest	Dominant sp.

Kavet ecological classes are commonly combined to allow for detailed description of ecological areas. But when Brao language users only require a basic area description for communication purposes, ecological classes are not generally combined. A good example of how ecological terms are linked is the term *bree baw-jeung-jun-dou*, which combines two base ecological terms, *bree baw* (a type of forest dominated by a particular bamboo species) and *bree jeung-jun-dou* (an area in the foothills of a mountain). Therefore, *bree baw-jeung-jun-dou* means a type of bamboo forest found in the foothills of a mountain. Another example is *bree lao-jarung-jarung*, which combines *bree lao* and *bree jarung-jarung*, and means a different kind of bamboo forest located on a very steep slope.

The Kavet's creative combination of ecological classification terms is not standardized, and is based on oral tradition. Therefore, Kavet do not always classify areas in the same ways, and individual knowledge of Kavet varies, as would be expected. However, Kavet rarely use contradictory terms to describe the same area. Instead, differences in classifications are frequently based on the level of importance put on particular habitat characteristics by individuals, and the scales that people are working with in different conversations. For example, if someone is concerned with the state of the soil in a particular area, that person will increase the use of soil classification terms. However, if soils are less important at a particular moment in time, then other terms may be emphasized. Therefore, ten different people could theoretically describe the same area using ten combinations of classification terms, and they might all be essentially correct. These findings indicate that there are actually many more than 108 classes of *bree*, and that there are almost an infinite number of ways to describe the same ecological area in Kavet.

Terms for vertical forest levels were not found in Kok Lak. However, they are implied. *Bree lip*, for example, is said to have a high abundance of vines and woody climbers in the upper canopy. *Bree kip grit* has a lot of woody climbers and vines in both the lower and upper parts of the canopy. Finally, *bree bok-doun* are short forests dominated by low vines. Because ecological areas tend to merge and exist in gradients,[51] they are described in Brao language. For example, the area *bree greung-deeak-bok-doun* essentially means the area straddling *bree greung* (old growth forest) and *bree bok-doun* (thick, low vine forest). *Deeak* means "edge."

ECOREGIONS

The Kavet also perceive broader ecological zones, or ecoregions, in which a large number of individual biotopes are found. During walks with Kavet, they often referred to "levels" when discussing broad ecological areas or ecoregions. The Kavet reported that they recognize four ecological levels in Kok Lak Commune, starting from the lowlands along the Sesan River and gradually moving up to the higher elevations along the Laos-Cambodia border.

The broad ecological zones are based mainly on altitude (the upper levels have generally higher elevations), and on the mixture of *bree* and *dak* habitats found in each ecoregion. The Kavet are easily able to identify which of the 108 biotope areas they recognize are found in each of their four broad ecological zones. Some biotopes are found in only one or two of the ecozones, although others are found in all four. It appears that these broad ecological classes provide the Kavet with a means for communicating broad ecological concepts. Interestingly, the Kavet hierarchical ecological classification system is not unlike systems ecologists use for classifying ecological regions.[52]

PLACE NAMES

Apart from classifying landscapes into ecological areas, indigenous peoples often put great importance on place names.[53] The Kavet consider place names extremely important, and names of places are more commonly used than ecological classification terms. Mountains, streams, places where events occurred, and former and present swidden fallows are the most significant Kavet named places.

SPIRITS AND BELIEFS IN NATURE

Spirits influence all aspects of the lives of most animist Kavet people, and many kinds are recognized by them. Spirits are believed to be more common and dangerous in deep forest areas far from human settlements. The two largest mountains in Kok Lak Commune, Haling and Halang, are similarly significant to the Kavet, who have strong taboos against hunting on the mountains. They only harvest a few plant resources from Haling-Halang, and then only after conducting special ceremonies to appease spirits, which include burning candles, praying, and smoking large tobacco cigarettes. According to Kavet consultants, only Kavet/Brao language can be spoken on these mountains. This is because the spirits might get angry as "they do not speak foreign languages." Moreover, the Kavet are not allowed to say bad things about other people when in these mountains, or complain of hunger.

A number of important ecological areas for the Kavet are associated with the occurrence of spirits. For example, *bree wang-houng*, *bree ga-wang*, and *bree gout-ja-naw* are all highly associated with spirits, and there are often restrictions associated with the use of these habitats. *Bree dom* was explained to us as being ecologically the same as *bree greung* (the oldest growth forests). However, *bree dom* are inhabited by some of the most powerful spirits, which prevent people from entering their forests or acting in certain ways when inside them. The Kavet believe that strangler figs (*Ficus sp.*), called *long jree* in Kavet, are the homes of dangerous spirits, and they sometimes cut these trees down to get rid of spirits that they suspect are causing illness.

KAVET LIVELIHOODS

Ample evidence exists that the Kavet have long inhabited and conducted swidden cultivation inside present-day VNP.[54] However, in 1993, VNP was formally created without any consultations with the Kavet. More recently, they have been restricted from living inside the park or conducting swidden inside it. This has resulted in serious challenges in livelihood, food security, and other areas.[55] Many of the Kavet from Kok Lak want to practice swidden agriculture in old, fallow *bree baw* areas, both inside and adjacent to VNP. The Kavet and other Brao subgroups have historically practiced swidden agriculture in low mountainous areas ranging from roughly 150 meters to 500–600 meters above sea level, and they have their own ways of organizing space to accommodate swidden agriculture.[56]

A high diversity of crops is grown in Kavet swidden plots. Baird et al. identified 181 different crop types that are regularly cultivated in Kavet and Kreung swiddens located in Kok Lak Commune, Veun Sai District, and Taveng Leu Commune, Taveng District.[57] These included 36 varieties of rice and 145 other types of annual and perennial crops. The average family cultivates between three and seven kinds of rice, and between sixty and one hundred types of other crops in swidden plots.[58] The Kavet also have a complex system of Animist practices that are integrated into the swidden agriculture cycle, and for the Kavet in Kok Lak, nine ceremonies are generally held over the course of a full swidden season.[59]

The Kavet conduct a different style of swidden agriculture compared to that of some other indigenous peoples. They tend to cultivate flat or slightly-to-moderately sloped areas, including valleys between mountains and the foothills of mountains. Many Mon-Khmer groups have similarly well-established rotating swidden systems.[60] During forest walks with the

Kavet in the upland areas within and adjacent to VNP, it was observed that most of the presently functioning swidden areas, as well as fallows, are situated in valleys. There are a number of reasons why the Kavet never cultivate the tops of mountains. First, they believe that the soils there are generally not suitable for swidden agriculture. Our Kavet consultants also reported that mountain areas used for swidden become excessively dry in the hot season, causing banana trees and other crops that normally survive the dry season to perish. In addition, mountainous areas are generally far from perennial streams and rivers, which are the main sources of Kavet drinking and domestic water, as well as being important fishing grounds. The Kavet generally do not want to carry water long distances to higher areas, or travel long distances to daily fishing grounds. Geography is important. Another reason for this is based on cultural norms; they have historically lived near streams, and the geography in their areas generally allows them to do that.

Situating swidden plots in the relative lowlands of mountainous areas has significant implications for reducing soil erosion and encouraging good regrowth during fallow periods. When the tops of mountains are left uncultivated, they act as important sources of seeds for reforesting fallow swiddens below. The Kavet carefully choose the places where they conduct swidden agriculture, and generally avoid cutting the largest trees in the forest, unless there are no other suitable areas available, as is often the case in lowland Kok Lak. Old forests are generally not cut because the labor required is much greater than for cutting down secondary growth. It is also much more dangerous to chop large trees. Other indigenous people, including the Brao in adjacent Attapeu Province, in southern Laos, also have a preference for shifting cultivation in secondary forests.[61] Kavet swidden plots are generally between one and three hectares. In Attapeu, the average size of swidden plots is about 1.5 hectares.[62] This is about the same average size as Kavet plots in Kok Lak, although people beginning to cultivate wet lowland rice generally have smaller swidden plots due to labor constraints.

The Kavet generally cultivate a piece of land for two years. All crops are grown during the first season, and in the second year both the size of the field and the crops grown are reduced. Short-term rice and sesame seeds are the most popular crops for second year swiddens. In the third year, swiddens are left for fallow, although bananas and small amounts of other crops may be harvested for years. Sometimes small numbers of perennial fruit trees are grown, such as mango and jackfruit, and the Kavet may return to harvest fruits from these trees for their entire lives, even for generations. In addition, fruit trees indicate where people previously did swidden, and are often referred to using the names of those who previously cultivated a particular area long ago. Fruit trees are important markers on the landscape.

The Kavet prefer to do swidden agriculture in *bree baw* bamboo forests because they are relatively easy to cut down, burn well, and produce fertile ash. They are often associated with high-quality soils, and they regrow quickly. *Bree baw* forests are often ready to cultivate again within three to seven years. Kavet frequently confirm that crops, including rice, grow much better in these areas than elsewhere, and that yields are high. The importance of *bree baw* has resulted in the Kavet categorizing it into more subclasses than other habitat types. *Bree baw* areas situated in slightly sloping areas along stream edges, and between streams, are generally considered the best areas for swidden. *Bree baw* areas in flat or slightly sloping areas, away from streams, are the second best for doing shifting cultivation, and the third are *bree baw* situated away from streams in moderately sloping areas.

The next most popular areas to do swidden are *bree lao* bamboo forests. However, sleeping in *bree lao* areas can result in itchiness caused by bamboo hairs, which is troublesome to the Kavet. The fifth most popular *bree* type for doing swidden agriculture are old secondary semi-evergreen forests, called *bree lawng* or *bree cha-chawng-lawng*. *Bree jyeul* wetland areas are also suitable for swidden agriculture, especially for growing sugar cane and bananas. However, they are generally not found in large tracts, and represent only a very small percentage of the total area under swidden cultivation. Other types of forests cultivated include the bamboo habitats called *bree pok, bree eung-le, bree ha-gou, bree ha-tiang,* and the vine and woody climber–dominated forests called *bree lip*. However, these areas are generally only cultivated when more favorable sites are not available. *Bree lip* areas are difficult and dangerous to cut down because of the high abundance of woody climbers in the upper canopy.

Soil quality is undoubtedly one of the most important factors affecting the choice of swidden areas, and even some *bree baw* forests with poor quality soils are not desirable for swidden agriculture. Areas with large amounts of small pebbles *(bree bree-o-bree-o)* or excessively sandy soils *(bree phaik-phaik)* are generally deemed unsuitable for cultivation. However, if there are some large boulders in the area, this is seen as advantageous, because plants cultivated next to large stones are cooler when temperatures are high and warmer when cool spells occur. The boulders essentially help regulate surrounding soil temperatures and prevent problems related to extreme temperatures. The presence of a certain number of large stones in fields probably also helps reduce soil erosion. However, areas with too many large boulders *(bree ta-maw)* are not considered good for swidden.

In addition, the Kavet rely on the presence of earthworm casts, called *bree ga-jouan-louan,* as an indicator of good quality soils. Dark-colored earthworm casts are linked to better-quality soil than whitish earthworm casts. The Kavet are reliant on the presence and absence of particular trees, vines, and herbaceous plants as indicators of the suitability of areas for swidden. For example, the occurrence of the herbaceous plant *teum grawdaw,* the vines *ja-meu ga-trawk* and *ja-meu chouk-bouk,* and the tree *lawng ja-ray* point to good soils, as does the presence of the herbaceous plant *teum treu-ay,* which is said to burn well and create fertile ash. On the other hand, the occurrence of *lawng ham* trees indicate acidic, poor-quality soils, which are not good for swidden. It is important to recognize that the Kavet do not cultivate all areas for swidden, and rely on a number of factors when choosing areas to cultivate.

SWIDDEN AGRICULTURE AND ECOLOGICAL SUCCESSION

Successional patterns are extremely important for swidden agriculture, both in terms of agricultural productivity and sustainability, and with regard to the maintenance of biodiversity values.[63] Swidden fields are called *meur* in Kavet. According to the Kavet biotope classification system, new fallows are called bree *char-char*, or simply *char-char* (young fallow). Fallow areas approximately three to ten years old, depending on the forest type and soil quality, are called *bree char-gra* (old fallows). The next stage is *bree char-chawng* (maturing fallow), which represents tall fallow generally suitable for cultivating. If the forest is semi-evergreen, the next stage of succession is *bree lawng* (tree forest), followed by *bree gra* (old forest), and finally *bree greung* (the oldest mature forest, with very large trees). If a forest is dominated by bamboo, the next stage after *bree char-chawng* is fully matured forest (see table 2). All of the above terms are commonly used in everyday Kavet conversation.

It is important to distinguish the successional patterns in the lowland ecoregion from the successional patterns in the mountainous or hilly ecoregions. The natural succession patterns in the extreme lowlands, near Kok Lak's villages, are remarkably different from those in the areas that are just a few hundred meters higher in elevation. In the lowlands, imperata grass, called *ta-ngaik* in Brao, is the main species that becomes established in fallows, along with the herbaceous annual *teum wai-meuang* and the nonnative thorny plant known as *Amerik* (*Mimosa* sp.). Moreover, the widespread occurrence of dry season fires set by ethnic Lao and Kavet people in the lowlands encourages the increased establishment of imperata grasses in fallows as time passes. The occurrence of fire has serious implications for successional patterns in the lowlands of Kok Lak. When imperata grass becomes well-established and fires are regularly set, fallow regrowth is greatly impeded, and many areas become climax imperata grass savanna areas (*bree ta-ngaik*). This is a change from the *bree lawng, bree gra,* or *bree kreung*–rich semi-evergreen forests that were historically found in the northern part of ecoregion number one.

Biodiversity values are certainly sacrificed when swidden agriculture is done in dense forests in the extreme lowlands. The Kavet are well aware of this problem and recognize that it would be preferable, from a successional and sustainability point of view, if they made their swiddens in ecoregion two, or further north. Kavet ecolocal knowledge about successional patterns in the extreme lowlands has caused them to be unhappy about being limited to cultivating lowland swidden plots. This sort of information is clearly very important for PA management, as it is key for understanding why keeping the Kavet in the lowlands has potentially more serious negative environmental repercussions than allowing them to live in upland areas in VNP.

The immense amount of knowledge that the Kavet have regarding successional patterns is shown by their ability to determine how old particular fallows are, even when they have not been cultivated for decades. When we traveled into ecoregion two, which begins in the low mountains about ten kilometers north of the Kavet villages, it was observed that successional patterns are markedly different from the lowlands, even though there is only a slight change in elevation. Fallows in the hilly areas have very little or no imperata grass growing in them. Instead, the herbaceous plant *teum wai-meuang* and the vines *ja-meu pra-leut* and *ja-meu glak-eel* generally grow up first, followed by either bamboo (generally *teum baw, teum lao,* or *teum eung-le* in ecoregion two), wild bananas (*teum graw-mat*), or fast-growing trees (mainly *lawng ch'ro, lawng long-lang,* and *lawng ha-lang*). The Kavet recognize that animals and birds also play important roles in determining what plants grow in fallows.

The pioneer tree species that often grow in fallow swiddens previously covered in *bree lawng* in ecoregion two generally become densely established in just one or two years, providing substantial protection from erosion. The Kavet explain that *lawng ch'ro* and *lawng long-lang,* the dominant species in most of the fallows observed in ecoregion two, grow up to ten meters tall and live for only five or six years, or until they are shaded out by other tree species. Despite their short life, they serve an important ecological purpose, because they provide much-needed shade for other long-lived plant and tree species that become established in their understory, and eventually overtake and shade them out. The shading function provided by just a few tree species makes it possible for a diverse array of plants and trees to become established, thus promoting greater biodiversity than is found in swidden fallows in ecoregion one, where these pioneers are rare. These successional patterns also benefit the swidden system, as fallow areas revert to dense forest quickly, making it possible for the Kavet to rotate back to them in a relatively short amount of time.

We asked Kavet consultants why *lawng ch'ro* and *lawng long lang* are so dominant in fallows in ecoregion two, when they are virtually absent in the extreme lowlands. They explained that the soils in ecoregion one are much sandier and of poorer quality, and are thus unsuitable for these species. They did not feel it would be possible to reseed lowland fallows with *lawng ch'ro* and *lawng long-lang.* However, they have never tried to cultivate these tree species, so maybe there is some potential. According to our Kavet consultants, the successional patterns in ecoregions three and four are unlike those in ecoregions one and two. In the latter, different pioneer species are found, and imperata grass or other seriously limiting grass or plant species are not present. Instead, the successional patterns there apparently result in good and fast regrowth, not unlike in ecoregion two. We have not yet been able to go to the northern ecoregions to observe successional patterns, and therefore cannot provide detailed descriptions of these areas.

FIRE MANAGEMENT

Fire invariably affects successional patterns in swidden areas, and systems of fire management are often especially important where shifting cultivation occurs.[64] Regular bush fires in the lowlands greatly impede regrowth in fallow swiddens in Kok Lak. The Kavet explained that they are unable to limit these fires because the lowlands are "Lao lands." The ethnic Lao living along the Sesan River have traditionally burnt the undergrowth of the *dipterocarp bree treng* forests near their villages, including areas close to the villages in Kok Lak. Fires are set to encourage the growth of young grasses suitable for livestock grazing and the growth of wild mushrooms, as well as to clear paths in the forest and to make hunting easier. Moreover, the occurrence of Kavet swidden plots in adjacent semi-evergreen forests over the last decade has resulted in the expansion of imperata grassland areas in the lowlands and, thus, the expansion of lowland burning.

The fire management situation in hilly "Kavet lands," however, is significantly different from that in the lowlands. The Kavet have strong traditional fire management rules in upland areas. Firebreaks are built around freshly cut swiddens in order to prevent swidden fires from spreading to surrounding forests. Moreover, local protocol prevents the setting of fires in forested or fallow areas, unless they are being prepared for agriculture. If anybody is found breaking this rule, they are required to pay a fine of one pig to the community. This effectively discourages indiscriminate burning, and we did not observe any signs of random fires in ecoregion two.

The only exception to this rule relates to the burning of undergrowth in *bree h'jawng* (malva nut tree or *Sterculia lychnophora*) habitat on mountain ranges. Each year, after collecting malva nuts from the forest floor, the undergrowth is burnt in order to keep the areas relatively open, making it easier for malva nuts to be collected in future years. However, swidden agriculture is almost never conducted in mountainous malva nut forests, partially because the thick root systems of malva nut trees make land preparation difficult. The Kavet said that they are able to manage fire in the upland areas because they only have to control themselves there, and do not have to deal with the ethnic Lao, who do not recognize the same traditions.

The growth of trees is defined using particular terms, apart from those associated with the agriculture cycle. If plants or trees have just broken the earth, they are sometimes collectively referred to as *but,* or grasses, even though everyone knows that not only grasses may be growing. Areas dominated by small trees about five centimeters or less in height are called *bree lawng hun-teum-houan.* When these trees are about thirty-centimeters tall, they are called *bree lawng hun-teum-gert.*

During our field investigations inside and outside of VNP, we encountered expansive areas of mature *bree baw* forests in the valleys and mountain foothills of ecoregion two, while the higher-elevation areas were generally covered with old-growth semi-evergreen forests (*bree greung* or *bree greung-jun-dou).* Since the *bree baw* areas almost all had a history of swidden cultivation, ranging from about twenty to almost one hundred years, the question of the influence that humans have had on altering habitat types is very relevant. This is especially true because there are only a few scattered large trees in *bree baw* forests, and we could not find any stumps of large trees.

Our Kavet consultants explained that the historical occurrence of swidden agriculture in the area has had an important influence on altering habitats. Although old-growth forests with no bamboo growing in them rarely grow back into *bree baw* bamboo forests, when *bree baw* bamboo is mixed with other tree species, swidden cultivation often results in the increased dominance of bamboo after the areas are left for fallow. However, they emphasized that it is difficult to generalize about the situation since soil types and topography also have a great influence on successional patterns. They claimed that even after decades, culturally influenced areas tended to continue to be dominated by *bree baw,* indicating that swidden-induced *bree baw* are climax habitats that do not easily revert to their former states. Nevertheless, the Kavet claimed that pure stands of *bree baw* also exist in areas apparently never subjected to swidden agriculture.

The Kavet do not see the successional situation in ecoregion two as being problematic, since old-growth areas in upland regions are not used for swidden, and remain covered in dense semi-evergreen forest. Moreover, *bree baw* forests permit a relatively sustainable swidden system, which is compatible with the overall livelihood systems of the Kavet. It is unclear whether Kavet agriculture patterns have reduced biodiversity, or whether they have encouraged a greater variety of forest habitat types that may actually benefit biodiversity—especially species of wildlife that require a variety of habitat types, including relatively open areas.

It is odd that the *bree baw* forests in the lower parts of ecoregion two appear to be immensely stunted. The Kavet claim that this stunting is natural and related to reducing elevations. Many *bree baw* areas along the Lalay Stream appear to be populated by only young bamboo regrowth, but they are actually populated by mature bamboo that never gets big. *Bree gar-yawng* is a term used by the Kavet to describe the natural occurrence of bamboo flowering, which leads to the dying of large stands of bamboo, and certainly has important successional implications. While the Kavet are well aware of this natural cycle, which they recognize as occurring periodically and with a number of bamboo species, they do not know how to predict the flowering of bamboo. However, they are aware that populations of rodents often increase dramatically shortly after an area of bamboo has flowered. Bamboo forests that flower generally cannot be used for swidden agriculture because the fallen and dying bamboo prevent the clearing of undergrowth, an important step in the swidden preparation process. Young bamboo that grow up as single shoots, rather than as clumps, after flowering events occur, are called *bree heu-toung.* Bamboo that die, but not due to flowering, are called *bree-baw-rum-brum.*

Successional patterns in ecoregion two have been influenced by the recent demise of elephants. The Kavet claim that elephants were found in the area until about 1990. In fact, we traveled along an old elephant trail during part of our trip. However, during the period when firearms were abundant and efforts were not yet in place to limit hunting, elephants disappeared. Our Kavet consultants often complained that while it used to be easy to move around in the forest by using elephant trails, it is now much more difficult to do so, due to the lack of trails and the occurrence of thicker vegetation. It is unclear what the long-term impacts of the loss of elephants will be on vegetation communities, but it seems likely that biodiversity will decline. However, one possible advantage of having less elephant trails is that important wildlife areas are now less accessible to large game hunters from the lowlands.

NATURAL RESOURCE MANAGEMENT

While local ecological knowledge (LEK) can include information passed down through generations, it is also generated through dynamic learning processes that come from direct experiences living close to nature. LEK is not a stagnant relic; it is as dynamic as nature itself.[65] LEK and the everyday experiences intertwined with it have certainly influenced the way the Kavet classify forest habitats. Of the 108 ecological biotopes identified by them, 57 are named after particular kinds of plants (see table 2). While not all plants have associated ecological areas, virtually all of the plants that the Kavet associate with ecological areas have some sort of utilitarian value. Topographical, successional, pedological, and other factors play essential roles in developing classification systems, but utility is probably the most crucial determinant. It is therefore reasonable to suggest that increasing our understanding of how ecological areas are delineated can provide crucial information concerning the utilization and management of ecosystems, including those in PAs.

In Kok Lak, at least one potential keystone habitat has been identified. *Bree jyeul* is a kind of spring-fed wetland forest that the Kavet classify based on the biotope's rich mineral soils (see table 2). While this habitat is relatively rare, it plays a critical role as a wildlife refuge, as it is one of the only habitats in VNP that is moist and cool during the hot season. The fresh green vegetation that flourishes in these ecological areas includes a diverse array of plants that both humans and animals like to eat. The Kavet sometimes use this keystone habitat for conducting swidden agriculture, and they have recently been obliged to convert most *bree jyeul* areas in ecoregion one to wet rice paddy fields. Should these rare and significant areas be devoted to agriculture or to maintaining ecological integrity? This management question might not have been identified unless local people had made us aware of it, but it is certainly worth considering in the future, especially in the context of promoting wet rice agriculture.

CONCLUSION

There is great potential for utilizing LEK in land use planning and natural resource management inside and adjacent to VNP. Certainly the Kavet have a deep understanding of the ecological processes that affect VNP, although their immense knowledge has so far largely remained untapped by PA staff. This is due to a lack of recognition regarding the knowledge held by the Kavet, and because language barriers have made it difficult for the Kavet to communicate their knowledge systems to outsiders. Addressing management issues identified as significant by indigenous peoples is more likely to be fruitful than trying to deal with management concepts and agendas that do not fit within the ecological classification systems of the locals who use the resources within PAs.[66] It is time to fully integrate LEK into PA management, and there may be no better place to do so than in VNP.

The Kavet have a complex and detailed ecological classification system. They are an intelligent people with a deep understanding of their environment and its natural processes. Unfortunately, government officials and outsiders have sometimes erroneously portrayed them as ignorant and a threat to the environment, despite the Kavet being largely supportive of protecting the natural resources and biodiversity values within VNP. The Kavet have repeatedly noted their willingness and ability to assist in conserving the resources included within and outside the PA from internal and external threats.

The Kavet have strong ecological, livelihood, and cultural arguments for wanting to live and conduct swidden agriculture in a limited part of VNP. They do not deny gradual change and the cultivation of some wet rice paddy in the lowlands, but they believe that swidden in *bree baw* bamboo forests should be part of their sustainable indigenous livelihoods. Thus, the Kavet would like to use a relatively small area of old, fallow bamboo forests inside and adjacent to VNP for swidden agriculture, indicating their recognition that some changes are inevitable. They recognize that the success of the appeal that they have been making for more than a decade will largely be predicated on their ability to be seen as being reasonable, willing to accommodate development policy, and respectful of the need to maintain biodiversity values within VNP. However, at present, the Kavet have only been allowed to establish community protected areas (CPAs) inside the park, and while some fishing and NTFP collection by the Kavet is allowed in these areas, swidden agriculture is not permitted, much to their dismay.

Finally, it is important for local people, whether they are indigenous or ethnic Khmer, to participate in natural resource management activities, including PA management, as a way of ensuring sustainable PA management at various levels, from policymaking to everyday monitoring and enforcement activities. With the Kavet already largely recognizing the necessity of future conservation, the main question is, how can they best and most ethically be involved in determining their own destiny?

NOTES

1. This study draws heavily on research conducted as a consultancy in 2000 for the local NGO, NTFP Project, in Ratanakiri Province. For additional information about the Brao/Kavet, and their interactions with Virachey National Park, see I. G. Baird, "Controlling the Margins: Nature Conservation and State Power in Northeastern Cambodia," in *Development and Dominion: Indigenous Peoples of Cambodia, Vietnam and Laos*, ed. Frédéric Bourdier (Bangkok: White Lotus Press, 2009), 215–48; W. Balee and A. Gely, "Managed Forest Concession in Amazonia: The Ka'apor Case," *Advances in Economic Botany* 7 (1989): 129–58; I. G. Baird, *Dipterocarpus Wood Resin Tenure, Management and Trade: Practices of the Brao in Northeast Cambodia* (Saarbrücken, Germany: Verlag Dr. Müller, 2009), 257; I. G. Baird, "Various Forms of Colonialism: The Social and Spatial Reorganisation of the Brao in Southern Laos and Northeastern Cambodia" (Ph.D. Dissertation, Geography Department, University of British Columbia, Vancouver, Canada, 2008); I. G. Baird, "The Case of the Brao: Revisiting Physical Borders, Ethnic Identities and Spatial and Social Organisation in the Hinterlands of Southern Laos and Northeastern Cambodia," in *Recherches Nouvelles sur le Laos*, ed. Y. Goudineau and M. Lorrillard (Etudes thématiques No. 18, EFEO, Paris and Vientiane, 2008), 595–620.

2. Brao is frequently spelled Brau, Brou and Preuv, depending on the author.

3. C. Keller, J. Jordi, K. Gregerson and I. G. Baird, "The Brao Dialects of Cambodia: Lexical and Phonological Variations," *Revue de l'Institut de la Langue Nationale de l'Académie Royale du Cambodge* (Institute of Language, Phnom Penh, *Special Issue*, July 2008), 87–152.

4. Baird, "Controlling the Margins, 2009"; Ironside, J. and I. G. Baird, *Wilderness and Cultural Landscape: Settlement, Agriculture, and Land and Resource Tenure in and adjacent to Virachey National Park, Northeast Cambodia*, Biodiversity and Protection Area Management Project, Ministry of Environment, Ban Lung, Ratanakiri, Cambodia, 2003.

5. Baird, "Various Forms of Colonialism."

6. Ibid.

7. S. Koy, "A Case Study of Virachey National Park, Cambodia," in *Indigenous Peoples and Protected Areas in South and Southeast Asia: From Principles to Practice*, ed. M. Colchester and C. Erni (Copenhagen: International Work Group for Indigenous Affairs, 1999), 134–51.

8. Keller et al., "The Brao Dialects of Cambodia."

9. Baird, "Various Forms of Colonialism."

10. D. Ashwell, "Cambodia's National System of Protected Areas," in *Cambodia: A National Biodiversity Prospectus* (Phnom Penh, Cambodia: IUCN – World Conservation Union, 1997), 60–70.

11. BPAMP [Biodiversity and Protected Area Management Project] 2003, "Virachey National Park Management Plan 2003–2007" (Ministry of Environment, Phnom Penh), 66 pp.

12. BPAMP, "Virachey National Park Management Plan."

13. Ibid.

14. Koy, "Case Study of Virachey National Park."

15. Baird, "Controlling the Margins."

16. Baird, "Various Forms of Colonialism"; Baird, "Controlling the Margins."

17. Baird, "Controlling the Margins."

18. Baird, "Controlling the Margins"; Ironside and Baird, *Wilderness and Cultural Landscape.*

19. Baird, "Various Forms of Colonialism"; Baird, "Case of the Brao."

20. Koy, "Case Study of Virachey National Park."

21. Baird, "Various Forms of Colonialism"; Baird, "Controlling the Margins"; Ironside and Baird, *Wilderness and Cultural Landscape.*

22. Baird, "Controlling the Margins."

23. Baird, "Various Forms of Colonialism"; Baird, "Controlling the Margins."

24. Baird, "Various Forms of Colonialism, 2008; Baird, "Controlling the Margins"; Ironside and Baird, *Wilderness and Cultural Landscape.*

25. Ibid.

26. Baird, "Various Forms of Colonialism"; Baird, "Controlling the Margins."

27. Baird, "Various Forms of Colonialism"; Baird "Controlling the Margins"; Baird, *Dipterocarpus,* 2009; Ironside and Baird, *Wilderness and Cultural Landscape*; BPAMP, "Management plan"; Koy, "Case Study of Virachey National Park"; I. G. Baird and P. Dearden, "Biodiversity Conservation and Resource Tenure Regimes – A Case Study from Northeast Cambodia," *Environmental Management* 32, no. 5 (2003): 541–50; I. G. Baird, K. Tuptim, and M. Baird, "The Kavet and the Kreung: Observations of Livelihoods and Natural Resources in Two Highlander Villages in the Districts of Veun Sai and Ta Veng, Ratanakiri Province, Cambodia"(Livelihoods and Natural Resources Study, Novib and Oxfam [UK and Ireland], Ratanakiri, 1996), 99; S. Chay, P. Oum, and C. Ros, "Current Status of the Virachey National Park, Ratanakiri Province" (Department of Nature Conservation and Protection, Ministry of Environment, Phnom Penh, 1995), 34; B. Emerson, comp., "The Natural Resources and Livelihood Study, Ratanakiri Province, NE Cambodia" (The Non-Timber Forest Products [NTFP] Project, Ratanakiri Province, 1997), 83.

28. F. L Dunn, "Rain-Forest Collectors and Traders: A Study of Resource Utilization in Modern and Ancient Malaya," *Monographs of the Malaysian Branch, Royal Asiatic Society* no. 5 (Kuala Lumpur, 1975), 25.

29. D. A. Posey, "Indigenous Ecological Knowledge and Development of the Amazon," in *The Dilemma of Amazonian Development,* ed. E. F. Moran, 225–57 (Boulder, Colorado: Westview Press, 1982), 228.

30. Ibid.

31. N. Barrera-Bassols and V. M. Toledo, "Ethnoecology of the Yucatec Maya: Symbolism, Knowledge and Management of Natural Resources," *Journal of Latin American Geography* 4, no. 1 (2005): 9.

32. R. B. Primack, *Essentials of Conservation Biology,* 2nd ed. (Sunderland, Massachusetts: Sinauer Associates, 1998), 35.

33. D. Meidinger and J. Pojar, eds., *Ecosystems of British Columbia* (Victoria: British Columbia Ministry of Forestry, 1991), 11.

34. C. O. Delang, "Deforestation in Northern Thailand: The Result of Hmong Farming Practices of Thai Development Strategies?" *Society and Natural Resources* 15 (2002): 483–501; G. J. Martin, *Ethnobotany: A Methods Manual* (New York: Chapman & Hall, 1995); D. G. Casagrande, "Conceptions of Primary Forest in a Tzeltal Maya Community: Implications for Conservation" *Human Organization* 63, no. 2 (2004): 189–202; E. F. Moran, "Limitations and Advances in Ecosystems Research," *The Ecosystem Concept in*

Anthropology, ed. E. F. Moran, AAAS Selected Symposium 92 (Boulder, Colorado: Westview Press, Inc., 1982), 3–32.

35. Ibid.

36. J. B. Alcorn, "Factors Influencing Botanical Resource Perception among the Huastec: Suggestions for Future Ethnobotanical Inquiry," *Journal of Ethnobiology* 1, no. 2 (1984): 221–30; Martin, *Ethnobotany;* Moran, "Limitations."

37. T. M. Li, *The Will to Improve: Governmentality, Development, and the Practice of Politics.* (Princeton, New Jersey: Duke University Press, 2007), 374 pp; R. P. Neumann, *Imposing Wilderness: Struggles over Livelihood and Nature Preservation in Africa* (Berkeley, Los Angeles, and London: University of California Press, 1998), 256 pp.; N. L. Peluso, "Coercing Conservation?" *Global Environmental Change* 3, no. 2 (1993): 199–217; K. Rao and C. Geisler, "The Social Consequences of Protected Areas Development for Resident Populations," *Society and Natural Resources* 3, no. 1 (1990): 19–32; P. Vandergeest, "Property Rights in Protected Areas: Obstacles to Community Involvement as a Solution in Thailand," *Environment Conservation* 23, no. 3 (1996): 259–68.

38. R. Steinmetz and R. Mather, "Impact of Karen Villages on the Fauna of Thung Yai Naresuan Wildlife Sanctuary: A Participatory Research Project," *Natural History Bulletin of the Siam Society* 44 (1996): 23–40; Moran, "Limitations"; Alcorn, "Factors."

39. G. B. Ingram, "The Need for Knowledge from Indigenous Communities in Planning Networks of Protected Habitat for the Conservation of Biological Diversity: Three Island Settings," in *Ethnobiology: Implications and Applications: Proceedings of the First International Congress of Ethnobiology,* ed. D. A. Posey (Belam, Brazil, 1990), 87–105.

40. L. Wester and S. Yongvanit, "Biological Diversity and Community Lore in Northeastern Thailand," *Journal of Ethnobiology* 15, no. 1 (1995): 71–87; R. E. Johannes, "Integrating Traditional Ecological Knowledge and Management with Environmental Impact Assessment," in *Traditional Ecological Knowledge: Concepts and Cases,* ed. J. T. Inglis (International Program on Traditional Ecological Knowledge and International Development Research Centre, Ottawa,1992), 33–39; F. Berkes, "Traditional Ecological Knowledge in Perspective," in *Traditional Ecological Knowledge: Concepts and Cases,* ed. J. T. Inglis (International Program on Traditional Ecological Knowledge and International Development Research Centre, Ottawa, 1992), 1–9; J. Sallenave, "Giving Traditional Ecological Knowledge Its Rightful Place in Environmental Impact Assessment," *Northern Perspectives* 22, no. 1 (1994): 16–19.

41. Martin, *Ethnobotany.*

42. Primack, *Essentials of Conservation Biology.*

43. W. Denevan, "Ecological Heterogeneity and Horizontal Zonation of Agriculture in the Amazon Floodplain," in *Frontier expansion in Amazonia,* ed. M. Schmink and C. H. Wood (Gainesville: University of Florida Press, 1984), 311–36; Primack, *Essentials of Conservation Biology.*

44. Balee and Gely, "Managed Forest Concession."

45. Moran, "Limitations"; Alcorn, "Factors."

46. R. Steinmetz, "Landscape Ecology and Wildlife Habitats: An Indigenous Karen Perspective in Thung Yai Naresuan Wildlife Sanctuary of Western Thailand" (Wildlife Fund Thailand, Bangkok, 1996), 88 pp.; Martin, *Ethnobotany.*

47. Steinmetz, "Impact of Karen Villages"; Martin, *Ethnobotany.*

48. Martin, *Ethnobotany;* H. Brody, *Maps and Dreams: Indians and the British Columbia Frontier* (Vancouver & Toronto, Canada: Douglas & McIntyre, 1981).

49. Ibid.; Steinmetz, "Impact of Karen Villages."

50. Ibid.

51. R. Steinmetz, "The Ecological Science of the Karen in Thung Yai Naresuan Wildlife Sanctuary, Western Thailand," in *Indigenous Peoples and Protected Areas in South and Southeast Asia: From Principles to Practice,* ed. M. Colchester and C. Erni (Copenhagen: International Work Group for Indigenous Affairs, 1999): 84–107.

52. R. G. Wright and J. M. Scott, "Evaluating the Ecological Suitability of Lands for Parks and Protected Areas Using Gap Analysis Databases," in *National Parks and Protected Areas: Their Role in Environmental Protection,* ed. R. G. Wright (London: Blackwell Sciences, 1995), 121–30; R. Thackway and I. Cresswell, "Toward a Systematic Approach for Identifying Gaps in the Australian System of Protected Areas," in *Ecosystem Monitoring and Protected Areas,* ed. T. B. Herman, S. Bondrup-Nielsen, J. H. M. Willison and N. W. P. Munro (Wolfville, Nova Scotia, Canada: Science and Management of Protected Areas Association, 1995), 473–83; Meidinger and Pojar, *Ecosystems of British Columbia.*

53. E. S. Hunn, *Nch'i-Wana "The Big River" Mid-Columbia Indians and their Land* (Seattle: University of Washington Press, 1991); K. H. Basso, *Wisdom Sits in Places* (Albuquerque: University of New Mexico Press, 1996).

54. Baird, "Various Forms of Colonialism"; Baird, "Controlling the Margins"; Ironside and Baird, *Wilderness and Cultural Landscape.*

55. Ibid.

56. Baird, "Case of the Brao."

57. Baird et al., "The Kavet."

58. Koy, "Case Study of Virachey National Park"; Baird et al., "The Kavet."

59. Emerson, "Natural Resources"; Baird et al. "The Kavet."

60. Y. Goudineau, ed. "Resettlement and Social Characteristics of New Villages: Basic Needs for Resettled Communities in the Lao PDR" (Main Report, Volume 1, Ostrom, UNESCO and UNDP, Vientiane, 1997); Y. Goudineau, ed., "Resettlement and Social Characteristics of New Villages: Basic Needs for Resettled Communities in the Lao PDR" (Provincial Reports, Volume 2, Ostrom, UNESCO and UNDP, Vientiane, 1997).

61. Ibid., Goudineau, "Resettlement and Social Characteristics," Volume 2.

62. Ibid.

63. K. Warner, *Shifting Cultivators: Local Technical Knowledge and Natural Resource Management in the Humid Tropics,* Forests, Trees and People, Community Forestry Note no. 8 (Food and Agriculture Organization of the United Nations, Rome, 1991), 80 pp.; Balee and Gely, "Managed Forest Concession."

64. Warner, *Shifting Cultivators;* H. C. Conklin, *Hanunoo Agriculture: A Report on an Integral System of Shifting Cultivation in the Philippines,* Forestry Development Paper no. 12 (Food and Agriculture Organization of the United Nations, Rome, 1957).

65. N. Haggan, B. Neis, and I. G. Baird, eds., *Fishers' Knowledge in Fisheries Science and Management,* Coastal Management Sourcebooks Series (Paris: UNESCO, 2007), 437; Berkes, "Traditional Ecological Knowledge"; Johannes, "Integrating Traditional Ecological Knowledge."

66. Steinmetz, "Landscape Ecology"; Steinmetz and Mather, "Impact of Karen Villages"; Martin, *Ethnobotany;* R. Steinmetz, "Participatory Biodiversity Surveying and Monitoring in Lao PDR: A Starting Point for Collaborative Management of Protected Areas" (World Wide Fund for Nature/Thailand Programme Office, Bangkok, 1999), 20.

CONCLUSION

CAMBODIA'S INITIATIVES TO LEGALLY AND FORMALLY ENGAGE forest-dependent peoples in the forest sector are still in the early stages of implementation. Nonetheless, the orientation of the National Community Forestry Program (NCFP) represents a sea change from earlier in the twentieth century when forest-dependent people were stigmatized and viewed as "illegal inhabitants" and "practitioners of slash and burn" encroaching on state forestlands. New partnerships between forest communities and the government that vest rural villages with management rights and responsibilities for state public forests also represent a very different approach in contrast to the leasing of large tracts of forests to timber concessions, a model that dominated Cambodia's forest sector throughout the 1990s. The long-term transition in the forest sector is something that will take decades to unfold. Cambodia has taken legal and policy actions that reflect a policy commitment to move forward quickly with the devolution of forest and fishing rights to rural villages, and it is already surpassing neighboring countries in the speed at which it is implementing the new NCFP. At the same time, the rapid leasing of state public forest lands to private sector firms under long-term Economic Land Concessions (ELCs) raises serious questions regarding the government's social, environmental, and economic priorities. The government faces complex choices about the future of the forest estate, with many stakeholders, the outcomes of which may shape the future of Cambodian society and economy.

In twenty years, will much of Cambodia's landscape be characterized by privately run rubber plantations, or will it be dominated by community-managed natural forest ecosystems? How will resource rights be determined? Can the needs for macro-economic development be balanced with strategies that support the resource rights and economic needs of rural communities and the sustainable management of the country's valuable forest ecosystems? Since 2005, it is estimated that 15% of Cambodia's total land area has been leased to private companies[1], most of that being state forest lands with resident communities. The intention of this transfer of management rights to private companies is based on their promise to develop the land and create jobs; however, evictions and conflicts with local families have increasingly drawn the attention of the media, human rights activities, donors, and the government. Protests in Phnom Penh by dislocated families are drawing growing public support in opposition to ELC policies. The Cambodian government has warned investors that it would revoke land grants from companies that failed to fulfill their development responsibilities. Most recently, Cambodian Prime Minister Hun Sen said the government would "temporarily suspend granting economic land concessions" to "ensure equity and to strengthen the effectiveness" of leased lands.[2]

As Chapter 1 notes, the government was unable to control timber concessions in the 1990s, received limited revenues from the logging companies, with little economic development created. Instead, social conflict and forest degradation was the more common outcome. ELCs may well have a positive role to play in Cambodia's economic development, but ELC issuance and management will need to be done in a transparent manner that is responsive to local needs to avoid repeating the timber concession experience. Equally important, priority should be given to the issuance of CF agreement in all areas with forest-dependent communities that are interested in participating in the NCFP. The Ministry of Environment also needs to move forward with its Community Protected Area program, engaging resident people in the management of national parks and wildlife areas, and protecting conservation areas from commercial exploitation. Ultimately, the Cambodian government needs to clarify its long-term vision and priority goals for the nation balancing social and ecological stability with economic growth.

There is a need to better evaluate and assess the trade-offs in balancing macro-economic growth with rural economic needs. The Cambodia Poverty Assessment of 2006 notes that in 1994, 47 percent of the population lived below the national poverty line, declining to 35 percent by 2004. Economic growth generated by the garment industry, construction, and other activities have disproportionately benefited urban families, while 91% of poor Cambodians live in rural areas.[3] This World Bank study concludes that "improving the rural economy will require...equitable access to common property resources as a critical source of income and security for the rural poor."[4] Consequently, there is a danger that continuing to lease large tracts of forests to companies to clear for sugar, rubber, and other estate crops will destabilize the local economy and create a growing number of dislocated, disenfranchised families.

National policy reforms and programs enabling devolution and decentralization of state forest lands management to communities are now in place in Cambodia as detailed in Special Section 1. These reforms were established through preemptive action taken by the central government,[5] with encouragement from the donor community, I/NGOs, and other components of the civil society. It is apparent that these macro-policies are frequently in competition with other macro-policies supporting economic development. It is also important to consider the political factors that shape decision-making, including the influence of powerful individuals with economic and political leverage, who use their connections for personal gain. This is certainly the case in Cambodia, where a small group of families have secured ELCs for large tracts of the country's forest lands.

Macro-policies that benefit low-income, rural communities, such as the NCFP, can only be implemented through a combination of political will and the presence of micro-policy reforms. In turn, these require new institutional changes and capacities that often require redefining roles and norms.[6] While macro-policies can be changed with a pen stroke, micro-policies, and the capacity to operationalize them, require time to develop and implement. In many cases, government and donors may have little leverage to effect such changes; rather it is an outcome of social process involving communities and civil society.

Now under way for nearly twenty years, this new community forestry movement in Cambodia has evolved from a handful of pilot projects in degraded forest areas in the Lower Mekong region, and implemented by small NGOs, to a national program with an elaborate legal and policy framework supported and driven by the Forestry Administration. Over the past five years, program implementation has accelerated, affecting hundreds of thousands of hectares and hundreds of forest-dependent communities, including large tracts of dense forest. This has been the result of a convergence of goals that link strong grassroots demand for greater forest tenure security with national government goals to conserve forests and ensure that natural resources contribute to the alleviation of rural poverty. This transition has also been accelerated in response to the rapid loss of forests due to logging and forest conversion to estate crops. The growing recognition by the government and many rural communities that retaining forests was an essential part of long-term national and local goals has driven reforms and effective social action.

Forest-dependent communities represent a natural partner for government forestry agencies and protected area managers. After timber concessionaires failed to demonstrate a capacity to manage state public forests sustainably, the Forestry Administration, in an effort to achieve the national goal of retaining 60 percent of the country under forests, has wisely sought the collaboration of the nation's forest communities. Community forestry has deep roots in Cambodia's cultural traditions, and has expanded in recent years as new concepts and structures have been introduced to its capacity to sustainably manage vast areas of the nation's forest ecosystems.

As discussed in chapter 2, community forest use areas may extend for 5 to 10 or more kilometers beyond their village, especially in regions with resin collection, with the result that communities currently utilize much of Cambodia's forest lands at varying levels of intensity.

Communities are increasingly motivated to take action to protect and manage their forests in the face of external threats. Chapter 3 highlights how 13 clusters composed of 58 communities in northwest Cambodia's Oddar Meanchey Province are working to protect nearly 70,000 hectares of forest lands that are under pressure from ELCs, migrants, illegal loggers, soldiers, and land speculators. These CF groups are working with the FA, NGOs, and Buddhist monks in creating an evolving partnership that may be able to restrain the powerful drivers of deforestation operating in their area.[7] The process of designing a community-centered REDD+ project has helped to establish these collaborative relationships, and though stakeholder interests may differ, there is a common goal of forest protection. The extent of which the project will successfully generate revenues for the communities and RGC remains to be seen.

Chapter 4 reports on the efforts of the community of Kompong Phluk to protect their flood forests since the 1930s, as they found agricultural expansion and resulting forest clearing was exposing the village to violent storms blowing off Tonle Sap Lake. As populations have grown and pressures on flood forests have increased, the community sought outside support to intensify its natural resource management systems, including both forests and fisheries. This involved developing management plans, zoning, designating firewood collection areas, and creating fish sanctuaries. A number of other communities around the Tonle Sap Lake have established management committees and use systems, with a growing number recognized by the Cambodian government under new laws, policies and programs, as discussed in chapter 1.

Chapter 4 raises the question whether these emerging community initiatives will be able to compete with other commercial interests, especially as the local economy, dependent on capture fisheries, experiences declining fish catches, climate change, and conflict with other user groups. The author concludes that communities are playing an important role in resource management that could grow, but requires external support, possibly in the form of payments for environmental services.

In chapter 5, three case studies from indigenous ethnic communities in Cambodia's northeast Ratanakiri Province illustrate how the Kreung village of Krala was able to build upon its traditional forest management systems with support from local NGOs, strengthening the capacity of the village to protect their family and community forest rights against land speculators; while other Tampeun and Kreung villages lost much of their forest and farmland to outsiders over the past twenty years. This experience indicates that indigenous ethnic minority communities can strengthen their management systems and resist external pressures to sell their forests, but often require external technical and organizational support to map their forests, develop management plans, and update resource rules and regulations to allow for some commercial activities such as cashew cultivation.

Chapter 6 provides a fascinating example of the role a group of Buddhist monks are playing in mobilizing communities to protect local forests from illegal logging and conversion to agriculture. The author observes that the monks and communities view forests as their "heritage, like Angkor Wat," and that it is their duty to protect them because it benefits the nation and their religion. The monks represent an ethical presence that is respected by government and communities alike, empowering them as effective advocates for forests and forest-dependent communities.

The monks have also been able to raise funds from local contributions and international sources to help finance forest protection efforts, representing a new model of forest management financing.

Rural communities retain a remarkable body of knowledge concerning the natural environments. The ethno ecology of the Kavet people, presented in chapter 7, illustrates the depth of local knowledge present in that cultural community. Generations of learning about medicinal plants and edible species, forest structure, wildlife behavior, hydrological patterns, and other natural phenomenon have created an impressive local knowledge base that can enrich emerging national forest management systems. More recent experiences in mobilizing community institutions and building their ability to formally govern forests, formulate long-term management plans, and interact with government and the private sector, reflect a new level of knowledge and capacity that can interface with traditional wisdom. For example, ethnic Kavet knowledge could inform and guide the management of Virachey National Park, both in monitoring and protecting the forests. At the same time, it would allow the Kavet to continue to reside and utilize some areas within the park and support a unique human eco-system to continue to be practiced.

As the dual growth engines of population and economic expansion accelerate pressures on forestlands, many communities see woodlands disappearing and degrading. They may feel the impact of environmental change on their microclimate, spring flows, and groundwater levels, as well as on the availability of forest foods such as tubers, edible leaves, fruits, mushrooms, honey, and other sources of household nutrition. Equally important, the loss of forest resources may impact family income. In some forest communities, over 50 percent of the cash generated by the family may come from the collection, processing, and marketing of a range of non-timber forest products. It is apparent that given the heavy reliance of forest communities on these environments, many are ready and willing to invest their time and resources in the protection of the forest and eager to find partners that can assist them to strengthen their tenure security.

The Ministry of Agriculture, Forestry and Fisheries (MAFF) and the Forestry Administration (FA) have made considerable progress over the past two decades in developing legislation and policies that protect the nation's forests. Capacities have also been built through the training and experience gained by Cambodia's new generation of environmental managers and professional foresters. Many of these individuals have studied abroad, worked with international agencies and NGOs, and spent considerable time in field research and management activities.

Cambodia's civil society has also become engaged in forest conservation and sustainable management. Buddhist monks are building conservation initiatives that draw on Cambodian values and traditions, and educating civil society and government on the need for effective environmental stewardship. Small NGOs are emerging in the provinces and districts to play important roles in training communities and building capacities that prepare them to participate in the National Community Forestry Program (NCFP). International NGOs have also played an important role in supporting and assisting these processes over the past twenty years, providing financial and technical support, as well as introducing community forestry innovations from other countries. Yet each group faces its own distinctive challenges as it strives to move Cambodia's forest sector transition forward.

Communities face numerous problems as they attempt to develop and formalize their forest management systems. Potential internal conflicts among village social groups and leaders may constrain the establishment of transparent and representative systems of forest governance at the community level. Forest communities are also challenged by external pressures, which range from individual migrants and soldiers that move into their areas to larger-scale concessionaires who may have backing from powerful actors within government. Some external pressures can be dealt with locally through community forest patrolling, boundary demarcation, and coordination with local commune councils and FA field staff; however, other pressures require support from the district and provincial government and national agency partners. While the multistep pathway to formal recognition of forest rights by the RGC requires considerable effort on the part of the community and its supporters, it may be the best way for securing tenure rights that are strong enough to resist powerful interest groups intent on gaining control of village forests.

Cambodia's community institutions were often disrupted during thirty years of civil war. Many villages were abandoned and later resettled. Traditional governance systems were disturbed and village boundaries confused. While Cambodian society has made considerable strides in re-establishing village and local government systems, the process remains ongoing. Establishing formal community forestry institutions and resource management systems is an important part of this process. Resource pressures will continue to grow in the future, and the clarity of forest governance and tenure security will be essential components for sustainable resource management.

The lack of technical capacity to map and monitor forests, develop governance and management systems, and interface with government is one of the larger challenges facing poor, rural communities. Training and extension programs and materials are being developed, but still are insufficient to meet the requirements of the NCFP. Strategies like training of trainers, village cross-visits, field workshops, and other capacity-building activities are effective, but require human resources and financial support. The development of district and provincial CF federations represents one way to develop long-term institutional capacity to support forest communities.

Finally, limited funding remains a major constraint to the development of community-based forest management systems. Forest villages tend to be some of the poorest in Cambodia due to their isolation and lack of capital. Without cash to invest in building management infrastructure and systems—including constructing patrolling huts, demarcating forest boundaries, purchasing flashlights and uniforms, as well as implementing fire control measures and assisted natural regeneration activities—CF groups are limited in their ability to halt deforestation and engage in sustainable management. Considerable success has been achieved by CF support projects that develop seed grant and small grant programs, which can provide modest amounts of capital to emerging community forestry management committees (CFMCs) in Cambodia. The need for capital will likely increase as more mature CF groups begin to implement management plans that require cash for investment in producing, processing, and marketing forest products.

The role of potential environmental service payments to forest communities and government agencies to conserve forests could provide some of the financing required. Chapter 3 describes experiences from the Oddar Meanchey REDD+ project, Cambodia's first community carbon offset initiative that was developed to assess the potential of REDD+ as a forest management financing strategy. The author concludes that while the project appears to have accelerated the approval of CF agreements and encouraged the government to restrict the issuance of ELCs in REDD+ community forests, the flow of funds from carbon sales has not yet started five years after the project was initiated. To a considerable extent, the slow start-up has been due to problems with methodology approval, global REDD+ certification systems, and international partnering agreements, as community and FA efforts to slow deforestation in the project area have been reasonably successful. Carbon sales may be forthcoming in the near future; however, the extent to which REDD+ funds will flow to communities remains to be seen.

The Forestry Administration is tasked with supporting and championing forest conservation and sustainable management in a context characterized by multiple, and frequently competing, development agendas. In Special Section 2, the author notes the difficulties government foresters face in advocating for forest conservation and sustainable management when there are other competing interests in government and the private sector. The Forestry Administration and the Ministry of the Environment must represent the needs of forest-dependent communities, and fulfill the mutual goals of sustaining forest ecosystems and responding to the economic needs of the people that live within and around the forest. In recent years, the NCFP has enjoyed support from MAFF and high-level officials within the RGC. This political commitment will need to be sustained to allow the program to be fully implemented in the coming decades.

Like its government forestry counterparts in other countries around the world, Cambodia's Forestry Administration is in the process of reorienting its professional culture from one geared towards timber production to one with a multifaceted approach that addresses a broad range of environmental and socioeconomic goals. Partnering with communities requires foresters to have skills in community organizing, conflict resolution, participatory mapping, participatory rural appraisal (PRA), and other tools and methods that were not part of the traditional foresters' training program. Attitudes may also need to be changed from a more custodial role to a support and extension-oriented function. Many forest departments find that some of their staff are resistant to making such changes in their roles. At an operational level, the Forestry Administration requires the capital resources and technical capacity to develop new skill sets needed for implementing the NCFP. Special Section 3 describes the eleven steps that prospective CF groups must follow to receive formal recognition, requiring considerable guidance from FA staff and local NGOs. While Cambodia has an increasing number of foresters within government and the NGO community who have the skills and experience to facilitate these processes, their capacity remains limited, given the urgent demand to extend CF control over state forests.

Inadequate national funds have imposed limits on salaries throughout the government. Cambodian forestry field staffs receive extremely modest salaries that make it difficult for officers to meet basic living requirements, often forcing them to seek salary supplements from other sources. This environment can create an institutional culture where corrupt practices take place, including forestry staff facilitating illegal logging, ignoring encroachment, and so forth. Efforts to eliminate corruption must continue to be made to create an environment where the NCFP can be implemented.

Over the next decade or two, as the NCFP is implemented, financial and technical support from the donor community will be important. International and Cambodian NGOs have provided much of the external support over the past two decades. More recently, the Japan International Cooperation Agency (JICA) and World Bank trust fund grants have been helpful in supporting capacity building. The larger shift toward full implementation of the NCFP will demand a more substantial commitment from bilateral and multilateral agencies, whose support must extend through early implementation of CF establishment and agreement formalization to the creation and implementation of site-specific management plans. If community forestry is to substantially reduce poverty in rural villages, it will need technical and financial support for viable enterprise development schemes. Initially, this may build on NTFPs and small forest livelihood activities, but in time will need to transition to higher income–generating commercial initiatives.

A past and ongoing problem with donors is the changing development priorities dictated by the home country. Forest sector management transitions, by their nature, require a minimum of several decades to achieve. Donors need to sustain their support commitments, awarding long-term grants and funding, as well as better coordinate their overall development strategies. The Technical Working Group on Forestry has been relatively successful in addressing the need for donor coordination, and this effort will need to be sustained in the future.

It has been popular for development policy makers to stress the role of the private sector as a financing vehicle for community forestry. The private sector has been identified, most recently, as a key buyer of carbon under emerging REDD markets. NTFP studies indicate that higher income streams can be established through improved production, processing, and marketing, requiring a stronger engagement with the private sector. The challenge for poor, remote, forest-dependent communities is how to gain that knowledge and develop equitable partnerships, given their lack of experience, knowledge, training, and capital. Unprepared, they will negotiate at a huge disadvantage. Individual and institutional intermediaries may be required to better organize CF groups and to provide support in training and networking to help them partner with the private sector, in terms of selling forest products, engaging in ecotourism, and marketing environmental services.

Community capacities lie in their commitment to, and knowledge of, place, their cultural ties to the forest, and their economic dependence on the forest for subsistence goods, cash, and environmental services. There remains, in many parts of Southeast Asia, considerable social capital reflected in the institutions, values, and practices of indigenous societies. On balance, rural communities and their local institutions are under pervasive and growing pressures from urbanization, privatization, modernization, commercialization, and individuation. All these forces have the capacity to erode indigenous culture and institutions that have played an important role in mediating community use of forests and natural resources in the past. Cambodia's experience is demonstrating that new partnerships can be created, which provide mutual benefits for forest-dependent communities, local governments, urban dwellers, and national planners. Cambodia's forests also possess considerable ecological capital. While Cambodia's rich natural forests have been degraded as a result of logging and other extractive activities, they still possess rich biodiversity and deliver important environmental and social services. Communities, with their local ecological knowledge are well positioned to manage and restore these important ecosystems.

Cambodia's national CF initiative is providing new opportunities to stabilize the nation's forest estate, a sector that was characterized as being under "chaotic" management during the 1990s, when national forestry agencies and conservation departments, due to their limited resources, were unable to cope on their own with the enormous extractive pressures bearing down on these vulnerable ecosystems. Cambodia has made marked progress in expanding the documented CF area from a few hundred hectares in 1998, to over 100,000 hectares with 143 sites in 2002, and 383,877 hectares by June 2010. The challenge facing all stakeholders is to move forward in expanding the area under CF in Cambodia, while also building the capacity of all participating communities and support organizations to manage their forests sustainably and productively.

Given the NCFP target of establishing one thousand CF groups that effectively manage two million hectares of forestland by 2020, attention will need to be given to creating awareness of community resource management programs, streamlining and simplifying the process, and exhibiting greater flexibility that will allow the integration of diverse local community approaches to management into the national programs, including those of Khmer villages as well as indigenous ethnic minorities.

Over six hundred additional CF sites need to be identified to meet the current NCFP goal of transitioning two million hectares of public forestlands to community management. Many of the CF sites that are currently part of the NCFP have been developed over the past ten to fifteen years with support from the FA, the FAO, and NGOs. The next generation of CF sites will require substantial external support and coordination with local government. Experience from Cambodia and other parts of Asia indicate that the most successful sites are those where communities are highly motivated and taking early initiatives to protect local forests.

Cambodia is currently at a crossroads in terms of its policy guiding the future of the nation's forests and rural communities. At present, village households are heavily dependent on natural forests in their area for food, fuel, building materials, fodder, biodiversity, medicinals, and NTFPs. The degradation of these forests threatens their livelihoods and their quality of life. Government policies are conflicting in supporting both devolution of forest rights to communities under the NCFP and other potential mechanisms including CPA and the Land Law, as well as allocation to the private sector for agro-industrial development. The RGC justifies ELC issuance as a means to stimulate economic growth and alleviate poverty, and while it may generate accelerated economic expansion, there is growing concern regarding lack of equity, environmental impact, and dislocation. One researcher notes that "land conflicts and forced evictions are already creating an environment of desperation and alienation, and leading to social instability in Cambodia…There seems to be a growing sense amongst the Cambodian population that the land problem is serious."[9]

Cambodia's forest sector transition is of immense importance to the country. After decades of rapid deforestation and forest degradation from unsustainable logging operations, the nation is now looking towards forest communities as one of the new management models for ensuring that 60 percent of the country remains under forest cover. Yet, the accelerated leasing of millions of hectares of forest lands to ELCs and mining concessions, including areas in the national protected area system, raises questions about the political will and commitment to achieve this goal.

Findings presented in Special Section 2 indicate that forests under community control are better conserved than those in protected areas. While achieving the nation's ambitious goal will not be easy, an alliance between forest villages and government agencies appears to hold the best prospects for meeting forest conservation objectives. A recent study of eighty community forests in Latin America, Africa, and Asia found that where communities held larger forest areas and had greater rule-making authority, they had better livelihood benefits and higher carbon storage.[10] In many nations, forests and forest communities face powerful drivers of deforestation that operate at a local, sub-national, national, and global level, involving a diverse array of actors. To slow and halt deforestation and forest degradation, it is necessary that there be an equitable and meaningful alliance between parties who are committed to the sustainable management of forests and who link communities with other parts of civil society, government and the private sector.[11] In many situations, only through addressing threats to forests that operate at differing levels, will effective management be possible. Communities will play a key role in such strategies, but without the backing of government and necessary technical support and financing, they will not be able to compete with politically and economically powerful interests. Reaching Cambodia's forest conservation goal will require a long-term commitment to policies and financing that support the recognition of community forest rights and respond to the economic needs of poor, rural villages. Cambodia's experience has relevance throughout the Asian region, where national poli-cymakers must decide whether they should exploit state forest resources for potential economic gain or devolve greater rights to rural communities to alleviate poverty and sustain critical ecosystems. Cambodia is giving increasing attention to the latter path, and will offer important lessons for neighboring countries facing similar forest choices.

NOTES

1. Mukul Devichand, "Has Cambodia Become a Country for Sale?" BBC News – Asia Pacific, January 12, 2011.

2. BBC News – Asia, "Cambodia Suspends New Land Grants for Companies," BBC, May 7, 2012.

3. World Bank, "Cambodia: Halving Poverty by 2015" (Phnom Penh, Cambodia: World Bank, 2006), ii.

4. Ibid.

5. David C. Korten, "Micro-Policy Reform-the Role of Private Voluntary Development Agencies," in *Community Management: Asian Experiences and Perspectives* (West Hartford, Conn: Kumarian Press, 1987), 309.

6. Ibid., 310.

7. Mark Poffenberger, "Cambodia's Forests and Climate Change: Mitigating Drivers of Deforestation," *Natural Resources Forum* 33 (2009): 285–96.

8. Ibid., 17.

9. Paul Vrieze and Kuch Naren, "Carving Up Cambodia One Concession at a Time," *The Cambodia Daily*, March 10–11, 2012, 7.

10. Ashwini Chhatre and Arun Agrawal, "Trade-offs and Synergies Between Carbon Storage and Livelihood Benefits from Forest Commons" *PNAS* 106, no. 42 (2009): 17667.

11. Mark Poffenberger, "Forest Communities and REDD Climate Initiatives," *Asia Pacific Issues,* no. 91 (Honolulu: East West Center, 2009).

APPENDIX

SPECIAL SECTION ON FOREST LAWS, POLICIES, AND PROGRAMS

ENABLING LEGAL AND POLICY ACTIONS

ROBERT B. OBERNDORF, J.D.

THIS CHAPTER PROVIDES THE READER WITH AN OVERVIEW OF the legal and policy framework that directly supports the successful implementation of community forestry in the Cambodian context. It is important to understand that without the supporting legal mandate, government and community stakeholders would not have the authority or the rights to implement community forestry arrangements, and that any implementation of community forestry in the country must comply with the legal framework as established. It is also important to have an understanding of how the framework supporting community forestry fits within the broader legal and policy framework relating to forestland resource management in the country, especially as the concept of community forestry expands into new areas in the country not previously considered in the current Forestry Law.

The primary legal instruments supporting community forestry implementation in Cambodia are the Forestry Law (2002), the Community Forestry Sub-Decree (2003), and the *Prakas* on Community Forestry (2006), though other legal instruments that provide the overall framework for forestland implementation in the country are also relevant for a full understanding of the Cambodian context. Various policies adopted by the

Royal Government of Cambodia (RGC) that support the implementation of community forestry are also relevant and discussed within this chapter.

It is important to keep in mind that many years of piloting community forestry through various government, donor, and civil society–supported projects took place before the legal framework that would ultimately support the implementation of formally recognized community forestry was completed. Lessons learned from the pilot activities were crucial for providing useful information that fed into the development of the existing legal framework. This process of learning continues today, and feeds into a continuously evolving legal and policy landscape supporting the successful implementation of community forestry in Cambodia.

OVERVIEW OF THE LEGAL SYSTEM

In order to understand the legal framework that supports community forestry in Cambodia, it is useful to understand the overall legal system supporting such a framework. The legal system in Cambodia exists within the overall governance structure that is created under the constitution, which is the supreme law of the land. The government is made up of a system where the intent is to have a clear separation of powers between the legislative, executive, and judicial branches of government. "The separation of powers ensures that no element or branch of government can assume absolute or dictatorial power, and it is a safeguard for the people against abuses of state power."[1]

The legislative branch is divided into the National Assembly and the Senate, with the authority to approve and amend legislation initiated by them or the RGC.[2] The executive branch consists of the Prime Minister, the Council of Ministers (or the Royal Government of Cambodia), and the various line ministries. The judicial branch consists of various courts including (but not limited to) provincial courts, a military court, and an Appeals and Supreme Court in Phnom Penh. The Ministry of Justice (MOJ) and the Chief Prosecutor implement criminal law and procedure, and oversee judicial police in the enforcement of all legislation through the courts.[3]

The laws and regulations of Cambodia are hierarchical in nature, and each of these derives its validity and authority from a rule placed above it in the hierarchical structure of the legal system. The following is a general outline of the hierarchy of the legal and regulatory system within the Kingdom of Cambodia:

1) Constitution

2) Laws (*Chhbab*)

3) Royal Decree (*Reach-Kret*)

4) Sub-Decree (*Anu-Kret*)

5) Decision (*SorChor Nor*)

6) Prakas

7) Circular (*Sarachor*)/Instruction (*SechdeiNanoam*)

8) Deika

Within this hierarchy of law are general observations that can be made on time and scope. The higher the level of the instrument that is being enacted, the greater the amount of time for actual enactment due to various levels of reviews it must go through. For example, a law that is drafted within a ministry must go through a review process within that ministry, the Council of Ministers (COM), the National Assembly, and, finally, the Senate, while a *Praka* is simply reviewed within the ministry that is promulgating it.[4]

Another aspect that should be considered is scope or reach of the legal document. Laws have broad scope and apply to all government entities and geographic locations within the country, unless specifically limited within their text. *Prakas* are only binding within the Ministry in which they are promulgated, and *Deika* only apply to the geographical area of the province, district, or commune that enacts them. These issues of time and scope are more fully explored in the detailed explanation of the different legal documents below.

Law: Laws, or *Chhbab*, are the primary source of law in Cambodia. *Chhbab* are the laws passed by the National Assembly (lower house) and the Senate (upper house). The *Chhbab* is often confused with the Kram, which is used for the promulgation of a *Chhbab* by the king or head of state. The overall process of drafting and enacting a law is a time-consuming process that can take years to complete.[5] For example, the recently enacted 2008 National Protected Areas (NPA) Law (2008) was initially drafted within the Ministry of Environment (MOE) in 2002, but was not finally enacted until six years later.

Royal Decree: The constitution states, "upon proposal by the Council of Ministers, the king (or head of state) shall sign decrees appointing, transferring, or ending the mission of high civil and military officials..."[6] Royal decrees have also been used in the past to establish nature reserves in Cambodia, though the constitutionality of such enactments were often questioned until ratified by the legislative branch in the recently enacted National Protected Area (NPA) Law.

Sub-Decree: Sub-decrees, or *Anu-Kret*, are regulatory enactments of the executive branch that are generally used to implement and clarify specific provisions within laws, though they are also utilized to outline the roles, duties, and responsibilities of government entities, such as a ministry, or for the appointment of high-ranking government officials. Sub-decrees tend to be drafted within a ministry or amongst several ministries that have subject matter competence on the area to be regulated. Once drafted, the sub-decree is submitted to the COM for examination and adoption. Once adopted by the COM, the prime minister signs the sub-decree. Authorization for the sub-decree, whether direct or indirect, must come from a higher-level legal instrument, such as a law passed by the legislative branch or the constitution. Since sub-decrees are adopted at the level of the COM/RGC, their scope can be quite broad and apply to all ministries within the RGC.

Decision: Decisions, or *SorChor Nor*, are issued by the cabinet of the COM or the COM/RGC on a specific issue, such as relating to a limited delegation of authority to a state agency, or granting of a special permission, such as granting a long-term lease over an area of state land to a legal entity in instances where there may not be established legal mechanisms by which to do so. Lower-level government entities, such as ministries, may also issue decisions on a specific issue that is within the scope of their subject matter jurisdiction.

Prakas: *Prakas* are ministerial or inter-ministerial regulations that are used, like sub-decrees, to implement and clarify specific provisions within higher-level legislative or regulatory documents. They are also often used for the creation of guidelines that are necessary for the implementation of laws or sub-decrees, such as the Community Forestry Sub-Decree. *Prakas* are usually drafted at the technical department level and then signed into effect by the minister (or ministers) in charge of the ministry within which the regulation or guidelines apply. It should be noted that *Prakas* are largely used to implement laws and sub-decrees in Cambodia, and that lessons learned can be quickly incorporated since the process for enacting or amending *Prakas* is relatively simple and quick. The drawback to *Prakas* is that their scope is limited to the subject matter jurisdiction of the ministries that enact them. In order to address this limitation, joint *Prakas* are often promulgated by multiple ministries and signed by multiple ministers.

Circulars: Circulars/instructions, or *Sarachor/SechdeiNanoam*, are instruments that are issued by the prime minister or a minister to explain or clarify certain legal or regulatory measures, or to provide instructions. Like *Prakas*, these are limited in scope, but easily issued.

Deika: *Deika* are orders given by local authorities (provincial, district, or commune) that have the force of law within the geographical and subject matter limits of their jurisdictional authority. *Deika* cannot conflict with or contradict existing rules and regulations enacted at the national level (laws, sub-decrees, *Prakas*, etc.).

A NOTE ON THE USE OF TECHNICAL MANUALS

Ministries or government agencies will often develop technical manuals that are used in the day-to-day implementation of delegated roles and responsibilities. These manuals are not legal instruments per se, but they are developed and used to help government staff understand and properly implement existing legal requirements. As such, it is important that the content of these manuals comply with provisions found in relevant legislation and subsidiary regulations, and will often incorporate or reference provisions found in relevant legal instruments. An example of this is the "Guidelines on Community Forestry Implementation" that was published by the Forestry Administration (FA) in 2006.

GOVERNMENT POLICIES AND COMMUNITY FORESTRY IMPLEMENTATION

The following is an overview of policies of the Royal Government of Cambodia (RGC) that are directly relevant to current community forestry implementation.

RECTANGULAR STRATEGY FOR GROWTH, EMPLOYMENT, EQUITY, AND EFFICIENCY—PHASE II

The Rectangular Strategy for Growth, Employment, Equity, and Efficiency—Phase II (2008) is the RCG's overarching socioeconomic development policy agenda for the Fourth Legislature of the National Assembly (2008–2013). The Rectangular Strategy is built on four fundamentals, including "ensuring environmental sustainability, especially through sustainable management and use of natural resources." It describes four growth areas that are prioritized by the RGC, including "enhancement of the agriculture sector," which covers (1) improving agricultural and diversification, (2) land reform and clearing of mines, (3) fisheries reform, and (4) forestry reform.

A NOTE ON EFFECTIVE POLICY DEVELOPMENT

The difference between policy and legislation is often confusing; in many instances commentators will refer to legislation, or even regulations, as policy, though they should be separate and distinct instruments in the governance field. Policy documents represent a course of action or vision that a government has adopted, written in a standard report format, while legislation sets out specific mandates, rights, responsibilities, and prohibitions on a subject matter within a rigid format of chapters and articles. Policy documents are written and adopted by a government's executive branch. Legislation and regulations are enacted by either the legislative or executive branches of government, with laws being enacted by the legislative branch and implementing rules and regulations enacted, implemented, and enforced by the executive branch.

Good policy documents should be clearly written and provide a government and its administrative agencies with clear directions on a course of action that is adopted. Policy documents are often adopted to direct the drafting, enactment, and implementation of legislation. Policy documents can also be written to assist in the interpretation of legislation by those entities that are responsible for implementation and enforcement. Properly written policy documents complement and link to legislative documents. In instances where there is no written policy on a subject matter, then the legislative documents are oftentimes referred to as the government's policy; this is considered a very poor form of policy development, and in such instances a clear policy should be written and adopted by the government.

For government policy to be truly effective, it should contain clear statements on courses of action to be taken by specific government actors that can be effectively measured and monitored over time. Far too often policy documents only contain broad statements embracing a general concept without spelling out clear courses of action by government actors.

Priority activities for the RGC in its fourth legislature include the following: accelerating land reform; public financial management reform; further implementation of the decentralization and deconcentration policy of transferring power from the national to subnational administrations; fisheries reform, including strengthening national resource conservation and taking serious action against illegal encroachment of flooded forests in order to secure fisheries resources; and forestry reform, including law enforcement, effective management of national protected areas (NPAs), climate change action, and community forestry. All development partners, including bilateral and multilateral development partners, private sector players, and nongovernmental organizations, as well as management and officials of ministries and institutions, are requested to assist the RGC in implementing the policies and programs outlined in the Rectangular Strategy.

NATIONAL STRATEGIC DEVELOPMENT PLAN UPDATE 2009–2013
The National Strategic Development Plan (NSDP) is intended to serve as the implementation tool or roadmap for implementation of the Rectangular Strategy. The NSDP 2006–2010 has, therefore, been updated to cover the years 2009–2013, the period of the Fourth Legislature and the Rectangular Strategy. It should be noted that this type of coordination

of national policy development and harmonization is an important step for the RGC, and the government should be encouraged to do the same with sectoral policies that are important for the sustainable management of forest resources in the country, such as the National Forest Program (NFP), which should be updated every five years in compliance with provisions found in the Forestry Law, and the National Protected Areas Strategic Management Plan (NPASMP), which should also be updated every five years in compliance with provisions found in the NPA Law (2008). Such coordination and harmonization of RGC policies will lead to more efficient use of limited human resources and better understanding and implementation throughout the country.

The updated NSDP sets a national target of achieving 60 percent forest cover, approving 450 community forests (noting that there were only 420 community forests at various stages of development as of 2010), and reducing fuelwood dependence by 2013. It also mentions the importance of the new NFP as the strategic framework for the forestry sector, and the role of protection forests, national protected areas, community forests, and improved management of forestry concessions towards achieving the national target of 60 percent forest cover.

Finally, the NSDP recognizes the importance of mobilizing resources, support, and financing to address the challenge of climate change, including REDD and greenhouse gas mitigation projects.

CAMBODIA MILLENNIUM DEVELOPMENT GOALS

Goal 7—ensure environmental sustainability—of the Cambodia Millennium Development Goals (CMDGs) sets out nine indicators for the forestry and environment sector under Target 13, "Integrate the principles of sustainable development into country policies and programs, and reverse the loss of environmental resources." These indicators and targets by 2015 include:

- *Indicator 7.1*: Forest cover: Baseline 60%, Target 60%

- *Indicator 7.2*: Protected Areas: Baseline 3.3 million ha, Target: 3.3 million ha

- *Indicator 7.3*: Protection Forest: Baseline: 1.35 million ha, Target: 1.35 million ha

- *Indicator 7.7*: Community Fisheries: Baseline 264, Target 589

- *Indicator 7.8*: Fisheries sanctuary: Baseline 264,000 ha, Target 581,000 ha

- *Indicator 7.9*: Fuelwood dependency: Baseline 92%, Target 52%

DECLARATION ON LAND POLICY (2009)

The RGC adopted a Declaration on Land Policy in July of 2009, which lays out broad principles and goals relating to land management principles in the country. While this policy does not mention the forestry sector specifically, or management of forest resources in the country, it does state that the process of state land registration (both state public and state private) should be accelerated, and it also calls for the development of a nationwide land information system (LIS) that is available for public reference.

As part of the land information system, the policy calls for the "establishment of a geology information system and soil classification based on natural characteristics of land," along with the development of a unified geographic information system across the country under the coordination of the Ministry of Land Management, Urban Planning, and Construction (MLMUPC). According to the policy, the MLMUPC will create a unit in charge of printing and distributing master maps, continue to install a geodetic network throughout the country, create a leveling network, and establish a permanent GPS base station.

The Declaration on Land Policy further specifies that state land trustee authorities, such as the Forestry Administration (the authority for the permanent forest reserve) or the MOE (the authority for national protected areas) will provide the MOEF (Ministry of Environment and Forests) with an annual inventory so that it may produce reports on the use of state land for the RGC. The policy indicates that the public can receive this information from relevant state institutions, and that State land trustee authorities, along with territorial authorities at all levels (provincial, district, commune), shall be responsible for protecting and ensuring accountability of public and private state land management.

As of 2010, work was under way to develop a comprehensive land policy in the country.

DRAFT SPATIAL PLANNING POLICY (2010)

A first draft of the Spatial Planning Policy was completed in 2010. While one cannot predict what this policy will eventually look like, there are already provisions that stress the importance of the sustainable management of natural resources in the country, along with general principles that link commune land use plans with district and provincial land use planning processes. It also links the land use plans of state land trustee authorities, such as the Forestry Administration's permanent forest reserves (PFRs) or the MOE's national protected areas (NPAs), thus linking and integrating top-down and bottom-up land use decision-making processes, and ensuring transparency and the participation of all stakeholders. Such a policy could impact the implementation of community forestry in the country, as plans for community forests are designed and implemented by various stakeholders.

NATIONAL FOREST SECTOR POLICY (2002)

This document adopted by the RGC prior to passage of the Forestry Law (2002) is not so much a policy as it is a government statement on broad management principles. It should be understood by the reader that the Forestry Law calls for the development of a national forest sector policy, which is to be developed by a national committee, chaired by the head of the Ministry of Agriculture, Forestry, and Fisheries (MAFF).[7] No timeline is established for updating this sectoral policy after it is initially approved by the RGC, as the Forestry Law only states that it may be revised if there are "significant changes in the conditions of the forest situation." It is likely that the RGC considers this document to meet the requirements for a national forest sector policy, as mandated by the legislative branch.

THE NATIONAL FOREST PROGRAM (2010–2029)

The Forestry Law states that the FA, "in compliance with the National Forest Sector Policy, shall prepare a national forest management plan with broad participation from authorities and concerned communities," which is to be approved by the RGC and updated every five years.[8] It appears that the current National Forest Program (NFP) meets the requirements for development of a national forest management plan, as required by the Forestry Law.

The NFP is probably the single most important sectoral policy document that currently exists for REDD+ implementation, and will continue to be so as it has been updated and modified over time to reflect current conditions in the Cambodian forest sector. During 2008 and 2009, the FA, in coordination with other stakeholders in the forest sector, drafted the NFP as a strategic framework, designed to guide the implementation of policy reforms mandated by the Rectangular Strategy and the National Strategic Development Plan.

The forest policy reforms prioritized under the NFP build on the new legal framework for forests that has been established, which is based on the 2002 Forestry Law; the Community Forestry Sub-Decree (#79/2003); community forestry guidelines (commonly referred to as the CF *Prakas*, enacted in 2006); various sub-decrees enacted to establish protection forest areas within the PFR; the Sub-Decree on Delineation, Demarcation and Registration of the Permanent Forest Estate (#53/2005); and ongoing reform of the country's forest concession management system.

The NFP, which was formally approved by MAFF in early 2010 and officially recognized by the prime minister, is being promoted by the RGC and development partners alike as a key policy document that provides clear guidance for the sector. One of the great strengths of the NFP is that it links and builds upon the legislative, regulatory, and governance frameworks already in existence, rather than trying to change that which has taken years of hard work to put into place. In addition, it clearly sets out actions that are to be taken by identified government actors over specific periods of time, thus making it a strong policy document that should be seen as a model for other sectors to follow.

The NFP identifies nine strategic priorities, including: the following contribution to the economy, climate change and REDD, forest governance, conservation of forest resources, improved forest management, and sustainable financing. The NFP prioritizes six programmatic areas that will receive emphasis over the next two decades in order to achieve these objectives, which includes the following:

1. *Forest Demarcation, Classification, and Registration.* Forestland is demarcated, classified, and registered by MAFF, and then entered on the land register by MLMUPC. The NFP sets a target of maintaining 60 percent forest cover by

2015, based on the CMDGs, with 120,000 kilometers of forest boundaries demarcated by 2029.

2. ***Forest Conservation and Development of Forest Resource and Biodiversity.*** The NFP sets targets of 3 million hectares of protection forests, 0.5 million hectares of plantations, and 2.4 million hectares managed according to sustainable forest management guidelines, with 50 percent of processed wood for export being certified and a chain of custody system established.

3. ***Forest Law Enforcement and Governance Program.*** This includes law enforcement and forest crime monitoring and reporting.

4. ***Community Forestry.*** The NFP sees local management as a key component of efforts to reduce deforestation and forest degradation and sets a national target of one thousand community forestry groups registered, covering two million hectares. The NFP also identifies the importance of broadening strategies for alternative models of forest management beyond community forestry (the Forestry Law and Community Forestry Sub-Decree limits the implementation of CF to production forest areas). It hopes to include community conservation forestry within areas designated as protection forest, and includes partnership forestry or community production forestry at larger scales than currently implemented in production forest areas.

5. ***Capacity and Research Development.*** The NFP is focused on building the institutional capacity of government actors responsible for forestland resources management and protection at all levels, and improving the ability of the government to carry out effective research in order to provide a sound basis for forestland resource management decision-making in the country.

6. ***Sustainable Forest Financing.*** The NFP identifies the United Nations' Reducing Emissions from Deforestation

and Forest Degradation (REDD) as a critical source of sustainable financing for implementation, and prioritizes development of national capacity to manage the proposed international REDD+ mechanism, including setting baselines and improving capacity for forest carbon monitoring. This is relevant to poverty alleviation, as it has already been shown that community forestry can provide an important platform for REDD implementation in the country.

FORESTRY LAW AND COMMUNITY FORESTRY

The Forestry Law (2002) is the primary piece of legislation that creates a jurisdictional and procedural framework for the management, development, use, and conservation of forest resources in Cambodia, and provides the legal basis for the existence of community forestry in Cambodia.[9] This law clearly states that MAFF has general jurisdictional management authority over forests, with one notable exception that the legislature clearly carved out (issues relating to the jurisdictional trustee authority over flooded forests is covered by the 2006 Fisheries Law):

> The state delegates management of nature protected areas to the Ministry of Environment, as stated in provisions of the Environmental Protection and Natural Resources Management Law of 24th December 1996, the Royal Decree on the Establishment and Designation of Nature Reserves on 1st November 1993, and other legislation.[10]

The Forestry Law also makes clear that the FA is the government authority within MAFF that has been delegated jurisdictional trustee authority over forest resources in accordance with the National Forest Sector Policy and the Forestry Law.[11] The FA is organized at the national level into seven separate departments with various roles and responsibilities relating to forestland resource management in the country, with the Department of Forest Management and Community Forestry being responsible for community forestry implementation.[12] At the subnational level, the FA is organized into four inspectorates, fifteen cantonments, and fifty-five divisions, which generally follow the jurisdictional administrative boundaries of local government territorial authorities in the country and helps to ensure proper coordination with these authorities on land resource management issues.[13]

The Forestry Law creates a nested state public forestland resource categorization system for management purposes, which recognizes and builds upon the state land management categories of the Land Law.[14] These categories include the permanent forest estate, permanent forest reserve, production forest, protection forest, and conversion forest.

Permanent Forest Estate (PFE)

The permanent forest estate is defined in the Forestry Law as encompassing the permanent forest reserve (state public land) and private forests ("forest plantation or trees, whether planted or naturally grown on private land under registration and legal title pursuant to authorized legislation and procedures").[15]

Permanent Forest Reserve (PFR)

The PFR comprises all state public forestland, except flooded forestlands managed under provisions in the Fisheries Law and forestlands in NPAs managed under provisions in the NPA Law. The PFR is further categorized as production forest, protection forest, and conversion forest.[16]

Production Forest

Areas within the PFR categorized as production forest are to be primarily managed and used for the sustainable production of timber and non-timber forest products, and can include areas under forest concession, forestland areas permitted for annual coupe production of forest products, degraded forestlands, areas of forestland reserved for forest regeneration or plantation, and forestland areas that have been granted to local communities to manage and benefit from under a CF agreement.[17]

Chapter 9 of the Forestry Law lays out the basic principles and rules for the minister of MAFF to grant areas of the permanent forest reserve to communities living in or close to the reserve for the purpose of establishing and managing CF in the country. While the law states that CF agreements are limited to fifteen years in duration, which is the standard time period for community natural resources co-management agreements, they can be renewed if the CF agreement and management plan are in compliance.[18] (Management plans are to be reviewed and updated at least every five years.) While communities have the right to harvest timber and non-timber forest products in the PFR for traditional use without the need for harvest permits or payment of royalties and premiums, communities are required to have permits issued and to pay royalties and premiums for any commercial harvests of forest products.[19] The law clearly states that "a local community cannot use the community forest in the form of a concession, nor sell, barter, or transfer its rights in such forest to a third party."[20] The Community Forestry Sub-Decree (2003), which is mandated by the Forestry Law, and technical guidelines outlined in the Prakas on community forestry (2006) provide the detailed rules and procedures for implementing CF in the country.

Protection Forest

According to Article 10 of the Forestry Law, "Protection forests shall be maintained primarily for protection of the forest ecosystems and natural resources therein," and can be established for the following: reserve forests for special ecosystems, research forests, forests for regulating water sources, forests for watershed protection, recreation forests, botanical gardens, and religious forests. It is the responsibility of the minister of MAFF to propose the designation of a protection forest in the PFR to the RGC, which shall be established through the enactment of a sub-decree.[21]

While there is no mechanism by which an area of protection forest could be granted to a community in the form of a CF in the Forestry Law or the Sub-Decree on Community Forestry (2003), communities may exercise traditional access and use rights within protection forests, as long as such traditional use activities only have a minor impact on the forest resources in question.[22] Under current provisions of the Forestry Law, if a community forestry–type arrangement were desired within an area designated as protection forest, then the forestlands proposed to be included in such an agreement would have to be reclassified by the RGC as production forest. An option to such an approach would be to amend the Forestry Law to allow for CF in protected forest areas, a process that could take years. The NFP has called for the creation of mechanisms to allow for CF types of arrangements in protection forest areas, or what are being referred to as community protection forests. How these mechanisms will be structured is not entirely clear at the moment, but it will probably be done through zoning arrangements incorporated into the required protection forest management plans.

THE COMMUNITY FORESTRY SUB-DECREE

The long-awaited Community Forestry Sub-Decree (CFSD) was passed by the Royal Government of Cambodia in 2003 after a lengthy process of public consultations and comment, which helped to establish a precedent for the use of such public consultations in the future. The passage of the CFSD represented an important "next step" in the effort by a variety of stakeholders who understood that community forestry could play an important role in sustainable land use and natural resources management, thereby leading to long-term, sustainable poverty reduction within the Kingdom of Cambodia.

The CFSD, with the general provisions on community forestry management found in the Forestry Law, creates a rather straightforward and easy-to-understand structure for implementation. However, many of the details required for full implementation are not included in the CFSD. These details have been included in the *Prakas* on community forestry that were finally enacted in 2006. In many ways, this is a good thing, since *Prakas* and technical guidelines are relatively easy to modify based on lessons learned during implementation, while the general structures within the relevant provisions of the Forestry Law and CFSD remain in place. *Prakas* and technical guidelines are frequently revised and updated in Cambodia.

PROCESS FOR ESTABLISHING COMMUNITY FORESTRY
UNDER THE SUB-DECREE

Though the CFSD states very clearly that the FA has the right to give official recognition of the demarcation of each CF boundary, there are two approaches to how this can take place.[23] The first is a bottom-up approach, where local residents living in or near the protected forest reserve (PFR) initiate the process in creating a CF community and request an area of the PFR as a community forest. The second is a top-down approach, where the FA will identify areas suitable for CF management, designate and demarcate areas of the PFR, and then work with the communities in the area to create CF communities and manage the resources accordingly. The area to be designated as a community forest, whether the process is initiated by the local community or at the national level, is based on what is essentially a situational needs analysis by the FA. This analysis is to be done with the involvement of local authorities, or commune councils.[24]

It is important to note that the FA makes the assessment for the designation of a CF area, but it is the duty of MAFF, through the minister, to issue a *Prakas* "recognizing and terminating areas requested for establishing community forestry by the Forest Administration." This authority for designation of an area as a community forest comes from specific language in the Forestry Law and is reiterated in the CFSD.[25] Once initial approval for the designation of an area as CF and the establishment of a CF community have been made, then the community needs to create a community forestry management committee (CFMC), which is created through secret balloting during a free and fair election by at least two-thirds of the members of a community during a public meeting.[26]

Once the CFMC is established, it is then responsible for drafting bylaws for its operation; a community forest agreement (CFA), which the FA and the community will enter into; and CF regulations that will control the use of the CF by the community and secondary users. The CFA will be a standard document that, once drafted, will be publicly posted for thirty days at the offices of the commune councils, district governor, and provincial governor. If no conflicts arise with the proposed agreement, then the Forestry Administration cantonment for the area in question can approve the agreement. It should be noted that the approval of the agreement is at the cantonment level, not the national level, of the FA.[27]

As stated in the Forestry Law, the CFA is in effect for a period no longer than fifteen years, but if the community forest is being managed in a sustainable manner in compliance with the CFA and community forest management plan (CFMP), then that agreement can be renewed for an additional fifteen years. Renewal can be denied if, based on monitoring and evaluation reports that were created with the participation of the CFMC, it is shown that the community forest is not being managed in compliance with the CFA, CFMP, and other rules and regulations.[28] In addition, the CFA can be terminated prior to the expiration date if there is written agreement among all parties to the agreement, the CFMC and at least two-thirds of the community members agree to terminate, there is serious noncompliance or violation of the conditions in the CFA or other provisions that leads to the nonsustainable use of the community forest, or the RGC decides that there is a higher public use for the area in question (compensation should be paid to the community if this occurs).[29] Once the community forest agreement is entered into, it is the responsibility of the CFMC to draft a management plan. The plan, once drafted, is sent for approval from the Forestry Administration cantonment level to the national level.[30]

OTHER PROVISIONS IN THE SUB-DECREE

In addition to the procedural provisions in the Community Forestry Sub-Decree (CFSD), there are also provisions that outline the basic roles, duties, and responsibilities of the communities, the CFMC, MAFF, and the FA.[31] These link with the roles and duties found in the Forestry Law and *Prakas* on the organization and functioning of the Forestry Administration. Roles and duties of local authorities and commune councils are not spelled out in a specific article, but it is clear from the test of the CFSD that they will only play a facilitation role during the establishment of community forests.

With regard to royalties and premiums, the CFSD reiterates provisions in the Forestry Law, but states that royalties and premiums required to be paid for commercial production "should be set after consultation with the Community Forest community in order to support community development, equitable benefit sharing, and poverty alleviation."[32] Another important provision that should be noted is the five-year moratorium on the harvesting of forest products (five years from the approval date of a CFMP). This provision applies only to timber products, not non-timber forest products. It appears that the new National Forest Program hopes to address the issue of commercial community forestry by establishing a separate classification with its own rules relating to production, management, and benefit sharing.

ENACTMENT OF THE COMMUNITY FORESTRY *PRAKAS*

The enactment of the community forestry *Prakas* in 2006 provided the final piece of legal framework necessary for the formal implementation and recognition of community forestry in Cambodia. The *Prakas* provided important details lacking in the Forestry Law and the Community Forestry Sub-Decree on the establishment of community forestry, including important new provisions relating to monitoring and benefit sharing. It also provided standard templates that could be utilized for the drafting of CFMC bylaws, community forestry regulations, the community forest agreement, and the community forest management plan. As lessons are learned during the implementation of the community forestry legal framework directly subsidiary to the Forestry Law, these community forestry *Prakas* will surely be updated and amended as necessary to reflect current knowledge and changing needs of the country.

COMMUNITY FORESTRY IN NATIONAL PROTECTED AREAS

The Ministry of Environment (MOE) and its secretariat for management of national protected areas (NPAs), the General Department of Administration for Nature Conservation and Protection (GDANCP), have been granted the jurisdictional trustee authority over NPAs in Cambodia.[33] While the currently existing NPA system in Cambodia was originally established and amended through the promulgation of various royal decrees, the legislature formally recognized and gave constitutional legitimacy to the previously established nature reserves through the enactment of the NPA Law (2008).[34] While recognizing the already existing NPA system in the country, Article 8 of the law also provides that the "establishment or modification of a nature protected area shall be determined by a sub-decree" after a proposal for such establishment or modification has been submitted by the MOE to the RGC, one which takes into account research and public consultations on the matter.

NPAs in Cambodia can be classified as national park, wildlife sanctuary, protected landscape, multiple use area, *Ramsar* site (a wetland protected under the Ramsar Convention), biosphere reserve, natural heritage site, or marine park; these classifications have yet to be defined by sub-decree as required under the law.[35] Article 11 of the NPA Law requires that every national protected area "shall be divided into four management zoning systems as follows":

1. **Core Zone:** Management area(s) of high conservation value, containing threatened and critically endangered species and fragile ecosystems.
   Access to the zone is prohibited except by GDANCP officials and researchers who, with prior permission from the MOE, conduct nature and scientific studies for the purpose of preservation and protection of biological resources and the natural environment. Also exempt are national security and defense sectors.

2. **Conservation Zone:** Management area(s) of high conservation value containing natural resources, ecosystems, watershed areas, and natural landscapes located adjacent to the core zone. Access to the zone is allowed only with prior consent of the GDANCP officials in the area, with the exception of national security and defense sectors. Small-scale community uses of non-timber forest products to support local ethnic minorities' livelihoods may be allowed under strict control, provided that they do not present serious adverse impacts on biodiversity within the zone.

3. **Sustainable Use Zone:** Management area(s) of high value for national economic development and management, including the conservation of the protected area(s) itself, thus contributing to the local community and the livelihoods of indigenous ethnic minorities. After consulting with relevant ministries and institutions, local authorities, and local communities in accordance with relevant laws and procedures, the RGC may permit development and investment activities in this zone in accordance with a request submitted by the MOE.

4. **Community Zone:** Management area(s) for socioeconomic development of the local communities and indigenous

ethnic minorities, which may contain existing residential
lands, paddy field, and field garden or swidden (*chamkar*).
Issuing land title or granting permission to use land in
this zone shall have prior agreement from the Ministry of
Environment in accordance with the Land Law.

Similar to the Forestry and Fisheries Laws, Chapter 6 of the NPA
Law lays out general rules and procedures authorizing areas located
within sustainable use zones to be granted to a community in the form of
a community nature protected area (CNPA). Areas can be granted for a
period of fifteen years, with renewal based on a community's compliance
with the CNPA agreement and management plans (management plans
must be reviewed and updated every three years).[36] In accordance with
Article 26, "communities and indigenous ethnic minorities shall not have
the rights to clear or work forestlands in the community protected areas
allocated to it, pursuant to the agreements with the Ministry of Environment, to practice agricultural farming or to claim title over the land, or to
sell, lease, pawn, donate, share, divide, or transfer the areas under its own
management to any person or legal entity." A draft CNPA *Prakas* had been
written and was going through a public consultation process in 2010.

INDIGENOUS COMMUNITY PRIVATE FORESTLANDS

Under the Land Law (2001), lands identified as being traditionally used
by indigenous communities in Cambodia can receive collective title.[37]
The Land Law essentially creates a unique situation for such indigenous
community title, which is also recognized in the Forestry Law (2002) and
clarified in the recently enacted Sub-Decree on Procedures for Registration of Land of Indigenous Communities (2009). It specifies that forestlands that have been found to be traditionally used by the community as
spirit forests, cemetery forests, or reserved land for swidden agriculture
(shifting cultivation), and which are classified as state public land, can be
included in the collective title and still be used for such purposes.[38] The
rights granted over these forestlands are limited in that the land cannot be
alienated (sold or encumbered by debt) and must continue to be used in
a traditional manner.[39] Essentially, the rights over land classified as State

public forestlands are granted to an indigenous community through a collective title and are limited by a deed restriction. If the deed restriction is not complied with, then the property ownership can revert back to the original property owner (the state).

The MAFF or MOE, depending on the location of the forestlands covered by the indigenous community collective title in question, would still have regulatory authority over the areas of state public forestland that is conveyed through the collective title (with the deed restrictions mentioned in place). In other words, these government institutions maintain regulatory authority to ensure that the forest areas are used in a traditional manner and that state interests are protected. The community has the possession, use, and management of land property rights transferred to them, with limitations (they can't do anything they want to with the land; they have to use it in the traditional way envisioned). This means that cemetery forest remains as cemetery forest, spirit forest remains as spirit forest, and the swidden/shifting/fallow areas remain managed as such, with the regulatory authority policing the deed restrictions. The community could not, for example, build a casino in an area that is intended as spirit forest.

Other than very clear limits placed on the amount of state public land that can be used as spirit forest or cemetery forest (seven-hectare limits for both),[40] it is not clear how much state public forestland the RGC will ultimately allow to be included in a collective title of an indigenous community. This will have to be determined on a case-by-case basis, with the coordinated involvement of the MLMUPC and the state public land trustee with jurisdictional regulatory authority over the area in question (MAFF/FA or MOE/GDANCP).

CONCLUSION

Much credit should be given to the RGC and other stakeholders in putting into place the necessary legal and policy framework that is currently supporting the successful and ever-expanding implementation of community forestry in Cambodia. It is truly one of the bright spots in the management of forest resources in the country and the region. There are still areas where further clarification and refinement are necessary, but this is part of an ongoing process of continual change and improvement. What is clear is that community forestry, as a natural resources management concept, has been fully embraced by Cambodia.

NOTES

1. Legal and Judicial Reform Policy, page 8, as adopted by the Council of Ministers (June 20, 2003).

2. The term legislation is being used narrowly in this chapter: legislation = laws passed by the legislative branch of government.

3. Constitution, Chapters 7-11h.

4. After promulgation, laws may also be reviewed for constitutionality by the Constitutional Council, but only upon request of the King, President of the Senate or National Assembly, the Prime Minister, one-fourth of members of the Senate, one-tenth of the members of the National Assembly or the Courts. The Constitutional Council has no authority to review laws on non-constitutional issues, nor does it have authority to review other legal instruments such as Sub-Decrees or *Prakas*.

5. This procedure for promulgating a Law would apply to proposed amendments to laws as well.

6. Constitution, Article 21.

7. Article 8, Forestry Law (2002).

8. Article 9, Forestry Law (2002).

9. Article 1, Forestry Law (2002).

10. Article 3, Forestry Law (2002); this article also states that the "management of flooded forests will be under a separate law."

11. Article 6, Forestry Law (2002).

12. Articles 5–12, Sub-Decree on Reorganization of MAFF (2008).

13. Annex, Sub-Decree on Reorganization of MAFF (2008), Article 6, Forestry Law (2002); *Prakas* on the organization and functions of the Forestry Administration (2003).

14. The process for delineation and demarcation of state public forestlands under the jurisdictional management and regulatory trustee authority of MAFF/FA is outlined in the Sub-Decree on Legal Procedures for the Establishment, Classification and Registration of the Permanent Forest Estates (2005).

15. Article 10 and Annex, Forestry Law (2002).

16. Articles 3, 10, and Annex, Forestry Law (2002).

17. Article 10 and Annex, Forestry Law (2002).

18. Articles 42 and 43, Forestry Law (2002).

19. Articles 24, 44, 52, and 53, Forestry Law (2002).

20. Article 44, Forestry Law (2002).

21. Article 22, Forestry Law (2002).

22. Article 10, Forestry Law (2002).

23. Article 3, CFSD (2003).

24. Article 7, CFSD (2003).

25. Article 23, CFSD (2003).

26. Article 17, CFSD (2003).

27. Article 26, CFSD (2003).

28. Article 27, CFSD (2003).

29. Article 28, CFSD (2003).

30. Chapter 7, CFSD (2003).

31. Articles 10,11, 21, and 23, CFSD (2003).

32. Article 13, CFSD (2003).

33. Article 4, NPA Law (2008), Article 3, Forestry Law (2002).

34. "This Law has a scope of application in protected areas defined by the provisions of the Law on Environmental Protection and Natural Resources Management, which was promulgated by Preah Reach Kram (Royal Decree) No NS/RKM/1296/36 of December 24, 1996, Royal Decree (Preah Reach Kret) on the Establishment and Designation of Protected Areas of November 1, 1993, Royal Decree on the Establishment and Management of BoengTonle Sap Biosphere reserve No NS/RKT/0401/070 of April 10, 2001, and other relevant standard documents." Article 2, NPA Law (2008).

35. Article 7, NPA Law (2008).

36. Articles 25 and 28, NPA Law (2008).

37. Articles 23-28, Land Law (2001).

38. Articles 16, 26, & 27, Land Law (2001); Article 37, Forestry Law (2002); Article 6, Sub-Decree on Procedures for Registration of Land of Indigenous Communities (2009).

39. Articles 26 & 27, Land Law (2001); Article 6, Sub-Decree on Procedures for Registration of Land of Indigenous Communities (2009).

40. Article 6, Sub-Decree on Procedures for Registration of Land of Indigenous Communities (2009).

COMMUNITY FORESTRY: A TOOL FOR SUSTAINABLE FOREST MANAGEMENT

ABU HASSAN AHMAD
ZAINAL ARIFFIN AHMAD
EDWARD MANINGO

SUSTAINABLE FOREST MANAGEMENT IS A CRITICAL ELEMENT for long-term development and poverty alleviation in Cambodia. Important prerequisites to sustainable socioeconomic development and poverty reduction in the country is the government adherence to the rule of law and principles of good governance, such as transparency, accountability, and public participation. Taking into account the welfare of the different stakeholders and sectors in the country and the international commitment to the conservation and sustainable management of forest resources, the Royal Government of Cambodia (RGC) passed the Community Forestry Sub-Decree in 2003.[1] The sub-decree aims to support poverty alleviation and decentralization by providing effective means for communities to participate in the reforestation, rehabilitation, and conservation of natural resources, forests, and wildlife. This led to the establishment of the National Community Forestry Program (NCFP), a component of the new National Forest Program (NFP).

The RGC also endeavors to implement harmonized sets of laws, programs, action plans, and institutional arrangements on forest resource development by promoting and encouraging public consultations and participation, particularly from local communities and concerned NGOs.

Unfortunately, progress is often complicated by the capacity of stakeholders and the conflicting roles and functions of many institutions. Competition among concerned departments within the ministries for turf and jurisdiction is triggered by demand for funding from donor agencies, as well as access and control over state public lands.[2]

In spite of these challenges, the RGC continues to pursue comprehensive forestry sector reforms designed to combat illegal activities and manage the forest resources of the country in ways that will ensure sustainability forest resources for present and future generations. The Forestry Administration (FA) has been mandated to implement this new direction, particularly to prevent and curb illegal forest activities, as well as improve the regulatory framework of the forestry sector. Cambodia has made significant progress in implementing these reforms and strengthening community capacity to participate in the forestry sector. This has become a cornerstone of the direction that sustainable forest management has taken in the country.

In Cambodia, the agricultural sector contributes 40 percent of the GDP. More than 70 percent of the total population is employed in the agricultural sector. The majority of the poor are heads of families who are employed in the agricultural sector and have little or no formal education.[3] Reliance on foreign assistance remains significant. While increasing donor support to forestry and environmental nongovernmental organizations (NGOs) resulted in strengthening community-oriented programs and building stronger management capacity, it has correspondingly reduced foreign and private sector investment in the forestry sector since 2001.

In Cambodia, the linkage between the rural people and the forest is very strong, owing to the high dependence of the community on the forest for their basic needs.[4] At the same time, the forests and forestlands in the country are under pressure from various sectors. Increasing exploitation puts more pressure on forests for goods and services, exacerbating rural poverty and worsening the conflicts over resources among various stakeholders and interest groups.

The problem of forest management is linked to rural poverty and overexploitation of the forest resources. A report by the World Bank, the "Cambodia Poverty Assessment," indicated that about 85 percent are dependent on farming and the forests for their basic needs.[5] Poverty widely contributes to natural resource and environmental degradation as poor households unwittingly degrade the environment in ways that may undermine the quality of their lives, as well as those of other members of the community.

FORESTRY REFORM AND GOVERNANCE IN CAMBODIA

Forestry reform is a key element in the RGC's Rectangular Strategy for Growth, Employment, Equity, and Efficiency. It is the role of the government forestry policy to ensure sustainable forest management and the use of forests to improve the livelihoods of people living in rural areas and to contribute to economic growth. Since forests are crucial for the livelihoods of the people, the RGC considers community forestry an important element in forest management. Accordingly, the RGC will continue to monitor and evaluate the effectiveness and efficiency of this program in terms of the improvement of livelihoods of rural people, the sustainability of forest resources, and the expansion of community forestry.

The Cambodian Millennium Development Goals aim to maintain forest cover at 60 percent by 2015. In order to fulfill its policy agenda, the RGC has been implementing comprehensive forestry sector reform since 1998, and has achieved remarkable results in establishing the foundation for improving sustainable forest management. With the passage of the current Forestry Law in 2002, the RGC has been fully committed to supporting community forestry through the development and implementation of a coherent policy, legal, regulatory, and institutional framework that recognizes and supports the establishment of formally recognized community forest sites throughout the country.

The Forestry Law defined the framework for management, harvesting, use, development, and conservation of the forests and the rehabilitation and restoration of degraded forests to enhance forest productivity and public functions. It was promulgated by the royal decree dated August 31, 2002. Its objective is to ensure the sustainable management of these forests for their social, economic, and environmental benefits, including conservation of biological diversity and cultural heritage. The RGC aims to achieve the following tasks: 1) forest resource conservation, 2) good governance, 3) socioeconomic development, and 4) poverty reduction. Each task contains the following activities:

1) Forest Resource Conservation

- Reclassify and dedicate the major part of remaining natural forest stands to ecosystems protection and biodiversity conservation functions;

- Promote conservation and protection strategies, such as protected forests, watershed management, genetic and wildlife resources conservation, ecotourism, and special management areas, with a maximum participation of the local population;

- Implement the strict application of the Code of Practice as a regulatory framework for the sustainable management of forest resources and forest concessions;

- Conduct extension, education, and public awareness campaigns at all levels of Cambodian society.

2) Good Governance

- Implement capacity building, institutional strengthening, and research programs at all levels;

- Conduct education, training, and public awareness campaigns, with particular regard to the participation of local populations within conservation and sustainable forest management plans;

- Establish a forest administration in which necessary steps of devolution of decision-making power can take place, and in which functional procedures for multi-institutional collaborations are grounded;

- Encourage, implement, and coordinate multistakeholder processes that enable the harmonization of different perceptions and objectives of the various forest-interest groups at local, regional, and international levels; and

- Promote transparent information for the forest sector.

3) Socioeconomic Development

- Promote high socioeconomic value of forest ecosystems through protection and biodiversity conservation;

- Promote the substitution of timber supplied from natural forest stands with timber produced in plantations by encouraging private investment and public participation; and

- Optimize the use, processing, and marketing system of forest products, especially plantation forest products, to support domestic demand and export markets.

4) Poverty Reduction
- Legally recognize and protect the traditional rights of local communities to use forest resources under the framework of food security and poverty reduction considerations;

- Optimize the benefits to local communities from the use and management of forest resources through the implementation of forestry and wildlife conservation concepts that are based on the participation of local populations.

Over the last decade, the RGC ratified the following sub-decrees:

- 2000—Sub-Decree on Forest Concession Management governing all forest harvesting activities and forest concessions;

- 2003—Sub-Decree on Community Forestry Management;

- 2005—Sub-Decree on Legal Procedures for Establishment, Registration, and Demarcation of Permanent Forest Estates to manage forestland and forest resources for sustainable development;

- 2006—Implementation Guidelines (*Prakas*) for Community Forestry to provide rules for the establishment, management, and use of community forests;

- 2008—Sub-Decree on rule for granting users' rights to plant trees within state forestlands.

Various action plans were formulated to achieve these objectives. The ongoing reforms in Cambodia have intensified government crackdown on illegal logging. As stated in its Statement on National Forestry Sector

Policy and National Strategic Development Plan, the RGC has been embarking on an intensive process of developing and implementing a National Forest Program (NFP) in 2010 aimed to place good governance and effective partnerships at the center of sustainable forest management. The NFP will address local, national, regional, and global needs by coordinating national and international partnerships to manage, use, protect, and regenerate forest resources for the benefit of present and future generations of Cambodians. As a framework for action and investment, the NFP also facilitates concerted and coordinated implementation of programs and activities by all stakeholders based on mutually agreed upon objectives and strategies.

The current NFP focuses on six implementation programs, constituting the key priorities of the RGC, namely: (1) forest demarcation, classification, and registration; (2) forest resources and biodiversity conservation and development; (3) Forestry Law enforcement and governance; (4) community forestry; (5) capacity and research development; and (6) sustainable forest financing. Achieving the goals stated in the NFP will take a long time as the policy reforms of RGC, particularly in the forestry sector, face many challenges, including operating efficiently with a meager budget and limited capabilities.

THE NATIONAL COMMUNITY FORESTRY PROGRAM

Community forestry (CF) was launched in Cambodia to address poverty in the rural areas. It is built on the long tradition of forest use by communities living in or adjacent to forest resources, and is recognized by RGC as an effective mechanism for sustainable forest management and livelihoods development. Since 2008, Community Forest Management Agreements (CFMAs) have been approved at an accelerating rate, with more nationwide expansion for CF planned. As of June 2010, there were approximately 380,976 ha of Cambodia's forestland either approved or in the pipeline for approval under the National Community Forestry Program.[6]

The National Community Forestry Program (NCFP) was launched in May 2006 and included as a key strategic function of the National Forest Program in early 2010. A major objective of the program is to arrange for official recognition of CF initiatives around the country, ultimately placing two million hectares of state forestlands under community stewardship. Other objectives refer to assisting these communities to improve the management of forest resources and ensure greater economic benefits from such management for all members of the community. The NCFP provides the policy framework for implementing community forestry initiatives and is based on the Forestry Law of 2002, giving authority to the Ministry of Agriculture, Forestry, and Fisheries (MAFF) to grant areas in the permanent forest estate to local communities in order to manage and benefit from the resources therein.

The NCFP is also included in the National Poverty Reduction Strategy (2002) that requires all sectors to contribute to the national goal of poverty reduction. In addition, the Cambodian Millennium Development Goal is to maintain 60 percent of the land area as forest cover. These policies have been incorporated into the Rectangular Strategy (2004), and further strengthened in the revised Rectangular Strategy (2008), whereby CF is prioritized as the principal vehicle for obtaining payments for carbon, through voluntary carbon markets and Reducing Emissions from Deforestation and Forest Degradation (REDD).

The legal framework for the NCFP was established in December 2003, with the approval of the Community Forestry Sub-Decree, which indicated a historic shift from informal CF projects to the emergence of CF as a national strategy that is formally recognized and embraced by the RGC. In 2006, the Implementation Guidelines (*Prakas*) on community forestry were signed, providing a basis for the formalization of CF. The *Prakas* describe the basic requirements and steps that communities can take for the establishment and implementation of CF. By mid-2010, there were 430 CF sites covering a total of 380,976 hectares, of which 212 CF sites (218,007 ha) have been approved by MAFF and 94 CF sites (113,544 ha) have signed CF agreements with the chief of the FA cantonment. Further discussion and more details regarding the implementation of the NCFP is discussed in Special Section 3.

FORESTS AND NATURAL RESOURCES

Cambodia enjoys relatively favorable conditions for development. It has plenty of flat agricultural terrain, productive freshwater fisheries, and a considerable stock of timber. The country has one of the largest percentages of forestland areas in the region, providing important habitats for many species of animals, birds, and plants. The forest cover in the country in 1965 was estimated at 73.04 percent of the total land area, but steadily declined to 59.82 percent in 1993 and 58.60 percent in 1997 (figure 1). In 2002, the forest area was 61.15 percent and decreased to 59.09 percent in 2006 (table 1).

Based on 2010 statistics, the forest cover is approximately 10.4 million hectares covering 57 percent of Cambodia's total land area (figure 2). In many rural areas, forest resources are heavily degraded or have been lost entirely. The loss of the forest resources constrained many communities' ability to obtain important subsistence and market products. The main types of forests in Cambodia are characterized as evergreen forests, mixed forests, deciduous forests, and other types (figure 3). The average total growth stock is 708 million cubic meters, and the volume per hectare is estimated at 66 cubic meters.

Evergreen forests are usually multistoried forests where trees maintain their leaves during the whole year. They comprise the lowland tropical rain forests, the hill evergreen forests, and the dry evergreen forest and areas along streams and rivers (gallery forests), composing nearly 27.70 percent (5,031,540 ha) of total land area. The mixed forests contain variable percentages of evergreen and deciduous trees, with the percentage of evergreen trees varying from 30 percent to 70 percent. Semi-evergreen forests continue to appear evergreen throughout the year. The dry-mixed deciduous and dry dipterocarp forests compose the deciduous forests of Cambodia. The deciduous forests drop their leaves more or less completely during the dry season. The deciduous forest covers 25.84 percent (4,692,098 ha) of the land area, and the remaining 5.55 percent (1,007,143 ha) are classified as other forest type.

Other forests include regenerating secondary forests, stunted forests, mangrove forests, inundated forests, and forest plantations. Regrowth of secondary forests is representative of a continuous, usually dense layer of smaller trees. Stunted forests grow very slowly because of poor site conditions on hydromorphic soils and rock outcrops. Heavily disturbed forests such as mosaics of forest, regrowth, and cropping, which correspond to shifting agriculture, make up more than 40 percent of the "other forests" category. Areas of old regrowth and young secondary forest in the process of regeneration after clear-cutting are also included in this category. Forestlands are further organized into four classifications: 1) protected areas (28 percent); 2) protection forest (14 percent); 3) forest concessions (30 percent); and, 4) other forests (28 percent).

Figure 1. Change in Forest Cover: 1965–2010

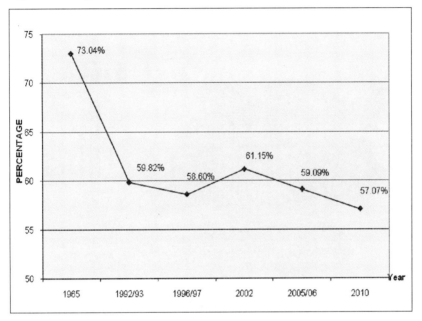

Source: FA

Table 1. Cambodian Forest Cover Estimates 1965–2010

YEAR	FOREST COVER	
	Ha	% of Total Forest Cover
1965	13,227,100	73.04
1992/93	10,859,695	59.82
1996/97	10,638,209	58.60
2002	11,104,293	61.15
2005/06	10,730,781	59.09
2010	10,363,789	57.07

Source: FA

Figure 2. Map of Forest Cover Change in Cambodia: 2002 to 2010

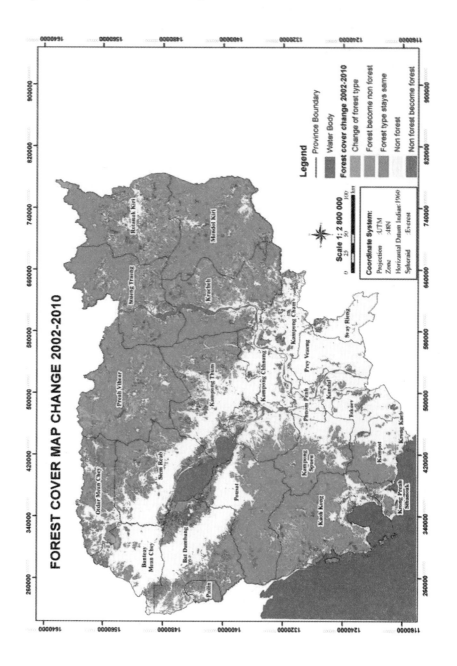

Figure 3. Forest Cover Map of Cambodia 2010

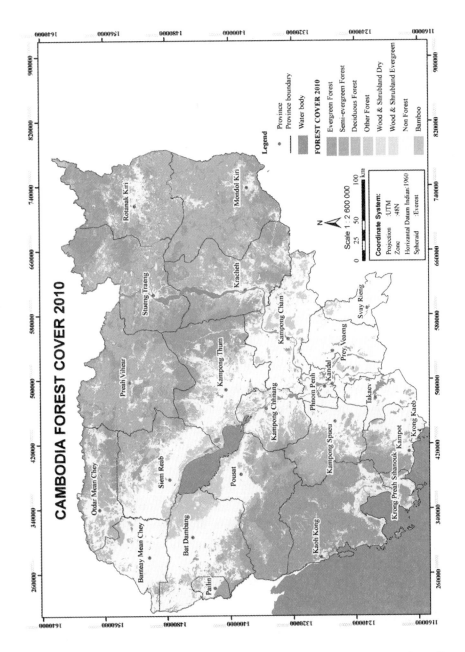

Source: FA

Figure 4. Forest Land Use of Cambodia

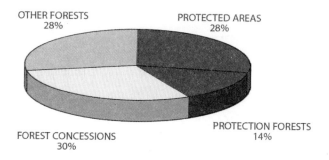

CAMBODIA FOREST LAND USE 2008

OTHER FORESTS
28%

PROTECTED AREAS
28%

FOREST CONCESSIONS
30%

PROTECTION FORESTS
14%

POVERTY IN FOREST COMMUNITIES

The rural economy of Cambodia depends on fish and rice, both of which are dependent on maintaining water flows that are regulated, in part, by forest cover. Fishing and foraging on common property such as lakes and forests regularly supplement the food supply. However, reports indicate that the economic condition of the rural poor deteriorates as access to natural resources declines.[7] Forest dwellers typically live in areas of low population density of less than eight persons per square kilometer. They are estimated to number about 450,000, or 3 percent of the population, and usually practice shifting cultivation in upland and mountainous areas, supplementing their agricultural income with forest produce, including timber and hunting.[8] Although the forest has contributed to the national revenue, it is estimated that the most the forest could contribute to the GDP is 10 percent. The country has great potential for tourism and is located in a fast-growing region, with land borders that allow it access to markets in Thailand and Vietnam.

A participatory poverty assessment identified the lack of food security as a major indicator of poverty in some rural forest areas. Food insecurity has many widespread effects. It influences health and nutritional standards and often forces villagers to have less concern for environmental considerations, and to use their natural resources in a non-sustainable manner. To address declining access to common property resources, such as fish and forests that are critical to the rural poor, the RGC has recently allocated all inland fishing lots to local communities for establishing Community Fisheries. Important economic and environmental problems in the villages are also related to the decrease in available agricultural land and falling soil quality.[9] The RGC has been implementing "Old Policy and New Action on land sector" by granting land certificates to local landless groups in total up to 2 million hectares for about 500,000 households.

Morbidity also plays a major role in sustaining poverty. Malaria, endemic in many parts of the country, is one of the major health problems in the rural areas. The disease has significant impact on socioeconomic development, negatively affecting both economic productivity and social well-being. There is a strong correlation between the prevalence of malaria and poverty, as poorer people are less able to prevent the spread of the disease or to seek cures. The high incidence indicates the difficulties of many people in protecting themselves, as well as the lack of adequate health services to treat the disease. Malaria is a leading disease in rural forest areas where mosquito habitats are found. In particular, the highest incidence of the disease can be found in the remote upland areas of northeastern, northern, and northwestern Cambodia, where it is difficult to reach people with prevention or treatment services. The RGC has targeted to end malaria by 2025.

Poverty has many consequences for communities, including deprivation or lack of income, education, poor health, and powerlessness.[10] It is important to note, however, that the classic measures of poverty, such as income and expenditure, have limitations in remote rural areas because many important aspects of livelihoods are not expressed in monetary terms, and some forest dwellers live in an almost cashless economy.[11] As a model of forest management, CF addresses these issues in a multivalent way within the forest communities by understanding that poverty should not only be based on income levels.

Poverty is multidimensional, and understanding the nature of poverty, particularly in the forest communities, is complicated by many factors, such as diversity of forest conditions, diversity of forest communities, and differences in the rights and opportunities to use these resources.[12] In Tonle Sap, poverty is widespread despite the vast natural wealth. The proportion of the population living in poverty in the five provinces surrounding Tonle Sap is the highest in the country, falling below the official poverty line.[13] The number of people living in the forests is numerically small, but the depth of poverty is greater, as is the challenge of lifting people out of poverty.

Forest communities typically lack the political power to formulate development policies, or the knowledge and resources to defend their rights to forest resources and prevent encroachment.[14] Usually, forest communities are located in remote areas with poor access to basic services, such as quality education and good health care. When the poor are forced into marginal forest areas, what is available to them may have little value. Lacking the capacity or rights to take advantage of resources, they may further degrade these resources instead of using them for livelihood development.[15]

The poverty in the uplands is aggravated by a rapid influx of poor migrants and by less development attraction to these remote areas, favoring private sector development investment and health programs in areas that are more populous.[16] Despite the importance of the forests as a source of livelihood to rural communities, access and productivity of these resources by the community have declined in past decades. The NFP seeks to reverse this trend with the NCFP, which is designed to reach the more remote and resource-dependent groups, enhance their tenure security, and provide new opportunities to raise their forest-based income.

POVERTY ALLEVIATION

In recent years, there has been a significant shift in conservation and natural resource management from greater state control to more community control, or Community-based Natural Resource Management (CBNRM). Among the important aspects of CBNRM is the participatory approach to project planning and implementation, ownership of the benefits of CBNRM by communities, access to land and associated resources as a key to self-determination, and capacity building. Building the capability of the community resource management organizations is viewed as a critical step in poverty reduction. A weak and poorly organized community often leads to weak and ineffectual management and enforcement activities.

In Cambodia, the NCFP is being implemented to channel financial benefits to the household economy. Community forestry has slowly been recognized at the central level. In 2002, about 64,000 hectares of community forests were identified by the FA, compared to the recent estimation of 380,976 hectares.[17] In the long term, it is believed that the economic benefits from CF will serve as a catalyst for economic growth in rural areas. People use forest resources to meet subsistence needs—such as wood, food, fiber, medicinal plants, and/or energy—as a "safety net" to meet occasional shortfalls in production or income. Among the CF members, the sustainable management of forest resources and tree planting when feasible can generate much-needed income, especially through the establishment of community-based enterprises.

Community forestry in Cambodia evolved as communities located in areas controlled by large-scale forest concessions were denied access to forest resources. The contribution of forests to poverty alleviation is widely recognized, especially in Cambodia where the livelihoods of the majority of rural Cambodians are dependent on access to forest products especially—for food, fuelwood, small-scale timber harvesting (for building and household materials), resin tapping, fodder, traditional medicines, and other non-timber forest products (NTFPs). The average per capita consumption of fuelwood for Cambodia's rural population is 455 kilograms per year, representing an average household demand for fuelwood of 2,730 kilograms per year. A household of six would spend up to $110 per year for fuelwood.[18]

Food security, income, nutrition, employment, energy sources, and overall community livelihood in rural areas are often linked with forests and other natural resources, and the capability of the forest to alleviate poverty is often thought to be determined by the condition of the forest. Good forests are thought to make people less poor and assist them in escaping from poverty, as well as preventing those on the margins from becoming poorer. The forest can also act as a "savings account" for people who invest their labor and other assets or as a safety net for the most vulnerable during times of hardship, such as droughts and agriculture price collapses. During hard times, poor people can harvest trees and other products to sell or to use in their own households.[19] Forestry activities, such as timber production and processing, are vital to the economies of many rural communities.

Income-generating activities in forest communities are primarily derived from NTFPs, particularly resin tapping and the collection of poles, vines, and rattan. In Ya Poey Community Forest, NTFPs collected in the village include bamboo and bamboo shoots, rattan, forest resins, tubers (cassava), wild taro used as pig food, wild pigs, deer, and wild chickens.[20] Fuelwood is the main (or only) source of energy for the majority of the rural poor. Most indigenous people, tribal communities, and "ethnic minorities" depend almost entirely on forests for their livelihoods. In some provinces, resin tapping from the forest provides communities with resources to obtain supplies of rice.[21]

Originally, community forestry was not designed to alleviate poverty. The focus was on rehabilitating degraded forest sites and protecting remaining forests, and the utilization of forest products was restricted to subsistence needs. Research on the relationship between forest dependence and community economic well-being is still inconclusive, as there is still a poor understanding of the relationship between the type and condition of a forest and the corresponding benefits to the local communities. For instance, although NTFPs in some forest areas have high economic value, the benefits are modest compared with royalties on timber received by the government. Little has been achieved in the forestry sector in terms of realizing the key role of forest resources in meeting the livelihood and subsistence needs of the nation's rural people. The NCFP will need to commit to developing supportive strategies that enhance the economic value of forests for rural communities.

As a result of recent policies to stimulate community-based forest management, increasing portions of forestland are falling under indigenous or community forest management and control. It is expected that these developments will create new opportunities. Community forestry has succeeded reasonably well in regenerating degraded forestlands, improving forest biodiversity, and enhancing landscapes and watersheds with little direct cost to national government. In addition, forest offenses gradually decrease after the establishment of community forestry, as the community becomes more aware of their roles and responsibilities as community forestry group members.[22]

Community forestry played an important role in the sustainable management of forest resources. Results from spatial analysis indicate deciduous forests comprise most CF areas. The deciduous forests with good forest cover are estimated at 32 percent, with only a small portion of the CF areas containing dense evergreen forests. Evergreen forests are considered more important to the community members than deciduous forests due to their diverse products, including NTFPs and fuelwood. Evergreen forests constitute about 20 percent of the forest cover, while only 4 percent are considered dense (table 2).

Table 2. Forest Cover in Community Forests

Forest Type	%
Evergreen	
▪ Dense	3.84
▪ Medium - Low Density	14.63
▪ Mosaic	1.83
Mixed	
▪ Dense	0.91
▪ Medium - Low Density	14.38
▪ Mosaic	0.59
Deciduous	
▪ Dense	32.42
▪ Mosaic	0.75
▪ Re-growth	5.12
Woodland/Shrubland	15.90
Others	9.63

Comparatively, the forest cover of CF sites is significantly higher than that outside the sites. Within a three-kilometer radius, the forest cover is significantly below 58 percent, while forest cover in the CF sites is estimated at more than 87 percent. The spatial analysis further showed that areas managed by the community have better forest cover compared to protected areas.

Looking at forest cover trends, it can be observed that the 2006–2010 decline in forest cover is higher (-3.10 percent) in the adjoining areas (within the three-kilometer radius of the CFs). This is a concrete demonstration of the effectiveness of CFs in arresting further degradation. It can be observed that in protected areas, the rate of deforestation has been higher than in CF-managed areas. While the decline in forest cover in the CF areas has decreased (from -1.07 percent to -1.02 percent), the decline of forest cover in protected areas has accelerated. These forests decreased by 1.52 percent from 2002 to 2006, and were further reduced by 1.75 percent in 2010.

Table 3. Change in Forest Cover

Location	Forest Cover (%)			Change in Forest Cover (%)	
	2002	2006	2010	2002–2006	2006–2010
Forests within the CF Areas	89.22 %	88.15 %	87.13%	-1.07 %	-1.02 %
Forests within 3 kms of the CF Areas	64.70 %	61.23 %	58.13 %	-3.48 %	-3.10 %
Forests within Protected Areas	81.99 %	80.47 %	78.72 %	-1.52 %	-1.75 %

Community forestry has made contributions in poverty alleviation, environmental sustainability (through forest conservation), and forest governance. CF has also proved to be a low-cost approach to forest protection that is useful in situations where the government lacks resources for enforcement. Despite the reported successes of CF, its full potential has yet to be tapped. Some recent analyses suggest that although CF has generally benefited local community members, it has yet to have a positive impact on the "poorest" of the poor.[23] The challenge now is to determine how CF might provide more income-generating activities and reach those in dire need.

Figure 5. Forest Cover Trend

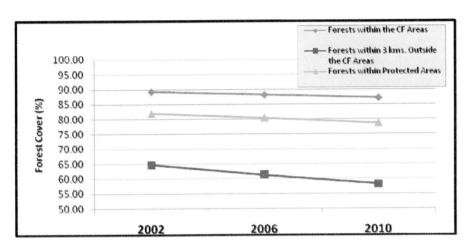

Most people in Cambodia are dependent on the forest in many ways, such as for fuel, food, bush meat, medicine, construction material, resin, and charcoal. In addition, CF could reduce the amount of time spent collecting fuelwood—time that could be spent on education, health clinics, leisure activities, livelihood generation, or other productive uses.

The goal of the RGC's forest sector strategy is to ensure sustainable forestry management, to utilize the forests to improve the livelihoods of people living in rural areas, and to contribute to economic growth. Community forestry provides good opportunities for rural development, since benefits from forest utilization can be directly transferred to local communities.

Villagers are concerned about the decrease of natural resources. Community participation is the key to successful conservation.[24] The willingness of different groups to participate in Community-based Natural Resource Management (CBNRM) can be influenced by how they perceive future benefits, including gaining stronger rights over the forests resources. On the other hand, community participation is sometimes constrained by factors such as poverty, education, access to training, and institutional capacity. Offering more immediate benefits (such as access to land for farming) can increase the motivation and interest of members. At the same time, farmers are unlikely to invest in costly improvements, such as terracing, unless they have secure land tenure that ensures benefits over the long run.[25] Furthermore, a CF provision stipulating that forest products be used only to meet a household's subsistence requirements is not an attractive proposition for the poor.

If local communities are to invest in forest resources or actively take part in conservation activities, they should be confident of their property rights. The success of community forestry is dependent on appropriate sharing of benefits with the community. The success factors for equitable benefit sharing depend on the participation and support of the stakeholders in decision-making, strong commitment from the local community, and clear roles and responsibilities from relevant stakeholders. Enabling community action includes recognizing rights and responsibilities of community groups, linking rights and responsibilities, and motivating markets.[26] Offering a favorable policy environment and improving capacity are important to generate the interest of community groups in collective action.

On the other hand, poverty will have an influence on participation in CF activities since it redirects the focus of households to more pressing needs. Securing food for daily sustenance is the immediate concern of most poor households, rather than participating in social commitments to conserve the forest. Understandably, this is because the benefits from the NCFP will likely be long-term in nature and, though sustainable, may not meet the immediate, short-term needs of the poor. Aside from poverty, health could also have a profound influence on a household's participation in community-based forestry activities. Specifically, health could affect labor productivity. Not only will it affect the household's participation in the community-based forest management activities, but it will also affect the poverty conditions of the household. Ultimately, the nonparticipation of poor households in community-based forest management may contribute to the degradation of forest resources. While forest degradation is in itself a result of poverty, it could be, at the same time, a factor that will determine the poverty and health of the community. Households may be disinterested in participating in CBNRM activities or engaging in sustainable resource management if their communities receive few benefits compared to other uses of the forest, such as timber harvesting. The lack of awareness of the importance of sustainably managing resources, as well as the appropriate management techniques, may limit community capacity to effectively manage their forests.

PAYMENTS FOR ENVIRONMENTAL/ECOSYSTEM SERVICES (PES)

The national grassroots movement in support of CF has demonstrated that local people are concerned about forests and are prepared to organize around forest protection and management. There is a need to encourage this initiative and support the involvement of local communities in forest management because of their strategic proximity to the resource. They are deemed to be in the best position to ensure effective management of the forest, and they are often very committed to conservation since they are most affected by forest degradation. At the same time, they are among the country's poorest people. By providing environmental services that benefit the nation, they should be compensated.

In sites where the NCFP is fully supported by the community or household members, positive impact can be expected. Aside from the impact of the program on the economic condition of household members, the active participation of the household is expected to promote positive values regarding the conservation of forest resources. It is then expected that there will be better management of the forest resources in the region. In Nepal, for example, community forestry user groups (CFUGs) have substantially contributed to an increase in forest cover.[27]

The emergence of CBNRM systems in Asia in recent decades was a reaction against a number of forces, including poor or unresponsive governance, unregulated timber exploitation, and the increasing disenfranchisement of local communities and associated inequity in the distribution of forest benefits. Deforestation and overexploitation of forest resources are often related to insecure tenure or unclear property rights. Property rights also play a central role in the establishment of PES schemes.[28] The extent to which the "poorest of the poor" were able to benefit from the program is unclear, mainly due to the lack of specific pro-poor community forestry strategies. Forest-related legislation that concentrates power in the hands of state governments still predominates in the management of CF sites in many countries.

In part, in response to the shortcomings of initial CBNRM programs, there has been a continuous improvement in community forestry systems in Asia; however, there is no single model for the development of social or community forestry. Most successful social forestry initiatives are those that most closely reflect their own social, cultural, and economic environments, while integrating basic principles and organizational guidelines, such as the devolution of natural resource management to local authorities, the right to self-determination, and user participation in management decision-making.

Community forestry groups often require assistance in finding ways to use forest products for purposes beyond their subsistence needs, thereby creating opportunities for generating non-farm income through community-based forest enterprises that have linkages to national and international markets. This often requires expansion of the community forest into higher-value forests. Commercial community forestry, a recent approach in CBNRM, needs to be piloted near high-value forest areas.[29] Creating appropriate business activities for the communities to make more income, researching appropriate markets for selling the community's products, and improving communication with national and international NGOs will help catalyze rural development in remote forest-dependent communities.

Community forestry groups face a challenge in developing institutional capacity and suitable operating procedures, although these are crucial elements for the successful implementation of community-based resource management programs. In Cambodia, the commune councils play a crucial role in community forestry development and the devolution of power that is necessary for effective implementation. Effective integration into the decentralization framework and identification of low-cost management models also determine the success of village resource management initiatives. The roles and responsibilities of government and its partners should be identified in developing, testing, and extending new approaches involving local benefit sharing. This becomes even more essential with the emergence of PES systems.

There is a pressing need to make community forestry contribute to the livelihood of village families. Some villagers put a low overall priority on environmental services compared to subsistence needs and the economic benefits of forests.[30] Viable income-generating options or strategies need to be pursued to improve the performance of CF. This includes extraction of NTFPs that provide relatively good income without degrading the forest resource base. For a community-based forest management system to be self-supporting, it must be financially viable. Operations and maintenance over time should be based on the revenues generated from CF group activities, which include NTFPs such as fuelwood, honey, fruit, and nuts, as well as from carbon credits. Sustainable environmental management and poverty alleviation are not mutually exclusive goals. The creation of markets for environmental services is one approach to increasing conservation and income generation for forest services.

Payment for environmental/ecosystem services (PES) is an evolving alternative funding mechanism that has the potential of sustaining CF development. PES systems are increasingly discussed as appropriate mechanisms for matching the demand for environmental services with suppliers, in this case community forest managers. There is a need to explore how payments for ecosystem services can be designed to generate funding flows that support sustainable resource management and livelihoods. The promise of environmental protection and economic efficiency has fueled the current trend in the forest sector to adopt market-based instruments rather than command and control systems of incentives.[31] Forests provide essential environmental services in addition to timber and other resources. Markets are viewed as useful and effective tools for environmental conservation.

The World Bank has been working to develop markets for environmental services in tropical forests, with a focus on the creation of markets for water. They also are working to establish market mechanisms for functions that conserve biodiversity and carbon sequestration. Another PES project approach is framed under the United Nations' initiative Reducing Emissions from Deforestation and Forest Degradation (REDD), which creates potential revenue streams from the sale of carbon credits representing avoided deforestation of natural forests and woodlots that would otherwise have occurred.

The basic principle behind PES is that resource users and communities that are in a position to provide environmental services should be compensated for the cost of their provision, and that those who benefit from these services should pay for them, thereby internalizing these benefits. PES will be a key requirement for sustainable environments and livelihoods. While also meeting environmental goals and objectives, markets for ecosystem services (MES) could also provide opportunities for poverty alleviation. There may be even greater opportunities by promoting PES projects to purchasers of voluntary carbon credits, but relaxing the stringent criteria outlined in the clean development mechanism (CDM) standards.[32]

CLIMATE CHANGE, REDD, AND COMMUNITY FORESTRY

Part of the government agenda is mobilizing resources to address the challenges of climate change. The RGC strongly supports the reduction of greenhouse gas (GHG) emissions from forests through conservation and avoided deforestation. Private markets are seen as useful and effective tools for environmental conservation, and for the conveyance of carbon credits through REDD and emerging voluntary markets. In the future, this may also be possible through the CDM. Both channels have the potential to provide opportunities to sustainably finance community forestry activities, improve livelihoods of forest-dependent communities, and act as a valuable mechanism for climate change mitigation.[33] The carbon financing models aim to bring 20 percent of the total forest area in Cambodia under community forestry management over the next fifteen years.

Since 2007, a REDD project initiative is being piloted in Oddar Meanchey Province to develop operational systems for REDD in Cambodia. Although the aggregate net annual rate of deforestation in Cambodia had declined to 0.5 percent during the period 2002–2010, the annual rate of deforestation in some provinces, such as Oddar Meanchey, remains high—estimated at 1.6 percent—making it a high-potential area for REDD. The initiative aims to provide long-term sustainable financing through the carbon markets to prevent forest degradation and deforestation.

Forestry activities, such as timber production and processing, are important economic activities in many rural communities. Yet only one-fourth of CF groups are located in good forests that are considered only slightly degraded or undisturbed. About two-thirds of all CF sites are situated in areas with either no or little forest, or heavily degraded forests. Although CF has the potential to address poverty issues and improve the management of the country's forest resources, research on the relationship between forest dependence and economic well-being of communities is inconclusive. The sustainability of CF groups remains uncertain, particularly given the limited capacity of the government to finance the development of CF institutions and activities.

The clean development mechanism (CDM), as well as private voluntary markets for carbon such as REDD, are viewed as potential sources of funding to sustain the operation of the NCFP. PES programs have been attracting increasing interest in Asia, which could be tapped as a viable financing alternative.[34] But while PES schemes exist in some developed countries and have been piloted in various locations in other developing countries, PES is still in its early stages of development and implementation. Many questions and issues remain regarding its design and implementation. In particular, the transaction costs for preparing CDM-compliant projects on such small scales are still too high. Continued research and testing are necessary to understand the potential and effectiveness of ecosystem markets.

Putting the theory of PES into practice is not an easy task. There is a need to build a body of knowledge and data on the role and impact of PES in poverty alleviation. It is argued that conservation should be the primary objective of PES and that focusing too much on poverty reduction may cause PES to become unviable, or to diminish the delivery of environmental services. Considering the market-based nature of PES, diverting the focus too much from environmental conservation could end up preventing the delivery of environmental services, and cause buyers to pull out and ultimately compromise the efficiency of PES. At the same time, there is a growing commitment by bilateral and multilateral agencies to see that newly emerging REDD+ initiatives include multiple environmental and socioeconomic benefits, including strengthening the forest tenure rights of rural communities. Cambodia's challenge is to explore PES systems as potential ways to finance the national forest sector transition and draw upon the important role that community forestry plays in that strategy. This approach may provide the key to conserving the nation's forests and enhancing rural livelihoods.

CONCLUSION AND RECOMMENDATIONS

After years of CF implementation in Cambodia, and also in other countries pioneering CF, the results are still varied. The outcomes of CF intervention are affected by the dynamically changing situation in the field, including the modality of government implementation. Moreover, endogenous factors such as malaria may have constrained socioeconomic development in some CF areas. Although CF is designed to support many livelihoods in rural areas, community forests are still dependent on external funding support and assistance. But growing opportunities in PES funding could aid the development of sustainable CFs. The evolving market for carbon sequestration provides an opportunity in CF areas. However, REDD opportunities largely depend on the capability of the community to negotiate the carbon market. As the experiences of CF vary from one country to another, there is a need to examine the CF experience in Cambodia to determine its contribution to poverty alleviation.

Community forestry in Cambodia has shown its contribution to improving sustainable forest management. While the rate of deforestation remains widespread in many parts of the country, the spatial data indicate that forest cover is significantly intact in areas covered by CF compared to adjoining areas. Community forestry has demonstrated its ability to reduce the rate of deforestation and improve forest cover. A more motivated and positive attitude of communities in CF areas has had a significant impact on reversing the trend of forest degradation. The positive attitude of the community members is critical to the sustainable development of forest resources. However, more efforts and assistance are needed to build the capacity of communities to manage their natural resources in a long-term sustainable and effective manner.

Another positive contribution of community forestry is that there is significantly lower poverty incidence among CF members than among those who are not CF members. While the data indicate that CF helps alleviate poverty, the results of the study have shown only limited contributions of the forest to livelihoods and household incomes. Nonetheless, despite limited income derived from the forest, the households attach a very high value to the forest resources they are protecting. This paradox could indicate possible non-income benefits of the forests. Most of the forests, which are still recovering, provide subsistence benefits, such as food, medicine, and fuelwood, among others. The community forests have not been commercially exploited, since these forests are mostly young and in the process of recovering.

Moreover, many of the communities have not yet prepared their CF management plans, which is a requisite for proceeding with commercial uses of the forests. The value of these benefits is often complex, but, nonetheless, will surely increase the total benefits that households derive. Furthermore, the measureable household income that can be easily discerned and revealed by the respondents will undoubtedly increase when CF management plans (CFMPs) are developed. Therefore, they can expect to engage in the commercial utilization of the community forests, and their household income should increase.

Strong participation among CF members is observably high. However, despite the collaboration among CF members, access to resources is still moderate. It can be observed that the protection of community forests, and the right to exclude outsiders from accessing the forests, is still limited. It is thus understandable that the community members expect outside support to strengthen their forest protection activities. Likewise, assistance is needed to educate and train the community forestry management committees (CFMCs) that oversee the forests.

REDD+ offers opportunities to attract alternative income to the communities. However, under present conditions of CF being relatively weak, the risks on the part of investors engaged in carbon trading through REDD is still high. This is significant in forests that have intense pressure on them from massive exploitation, land speculation/encroachment, and the high dependence of the community on wood resources to support cooking and housing. It is imperative that the drivers of deforestation be controlled for the carbon market to succeed. This will also require that the capacity of the CFs to enforce forest laws should be strengthened. The success of any REDD project will also depend on the communities, who are the primary stakeholders, receiving the major share of the proceeds from the sale of carbon credits.

The implementation of CF in Cambodia has brought significant outcomes in the sustainable management of the forest. Increasing access to forest resources by the community has been proven to effectively improve the economic condition of the community. The involvement of the community in the management of the forest also proves to be an effective method for mitigating poverty. But despite the success demonstrated by CF in Cambodia, much has to be done to further its implementation. Specifically, several policies and programs are needed to alleviate poverty in the field.

IMPROVING THE CONDITION OF THE FOREST RESOURCES

While community forests are deemed very important among the rural community members, they have made only limited contributions to household incomes. Apparently, this is partly caused by the poor and degraded conditions of the forest resources. It is imperative that existing forest stands be improved before the CFs can significantly contribute to household income. The community should exert efforts to protect the forest from outsiders, and rehabilitation measures through enrichment planting are also needed. The sustainable extraction of NTFPs should be observed by applying conservation measures.

STRENGTHENING PROPERTY RIGHTS, TENURE, AND ACCESS TO RESOURCES AND BENEFITS

Community forestry must be expanded to cover those areas that remain as open access. This will prevent further encroachment by outsiders.

STRENGTHEN LIVELIHOODS AND SUSTAINABLE MANAGEMENT OF THE COMMUNITY FOREST

Cottage industries need to be promoted in the rural areas. The livelihoods of the forest-dependent communities should be strengthened since many community members who depend on the extraction and utilization of forest resources realized only marginal incomes, while the middlemen captured the highest profit margins. Considering that some communities are not well-endowed with natural resources but have skills in processing, forest-based livelihood industries can be pursued by integrating key industry players. Production needs to be enhanced, possibly through agro-forestry technologies. Upon recovery of the forest stands, harvesting of forest products may be done on a sustainable basis.

CAPABILITY BUILDING

Contiguous forestry capability building is needed in the rural areas. Technical assistance should be provided to the different communities in management and governance of the community forestry. Monitoring is needed to spot communities who lag behind in terms of development. The capacity building needs to focus on the protection as well as the governance of the CFs.

INCREASING COMMUNITY FORESTRY PARTICIPATION

Community participation should be encouraged. However, to encourage greater CF participation, it is necessary that an enabling environment is in place. This includes favorable policies that encourage communities to invest in forestry projects; fair and equitable sharing of benefits, especially in the matter of carbon markets; greater transparency in the governance of CF projects; and incentives and technical assistance to the communities.

PROMOTING ENTERPRISE DEVELOPMENT AND VALUE ADDING OF FOREST PRODUCTS

Cottage-level industries and value adding must be pursued. Market linkage is essential for this to succeed. The community should endeavor to supply enough forest products to meet the increasing demand. The community, however, should be assured that it receives a fair market price for the forest products sold, whether raw or processed.

ACCESSING ALTERNATE FUNDING FROM REDD

The carbon market offers a potential source of funds for community forestry. However, the scheme is generally complicated and needs the participation of different key players. A clear framework for developing the market for environmental services is necessary for this to succeed.

ADDRESSING MALARIA-RELATED HEALTH PROBLEMS

Malaria remains a problem that constrains the economic development in CF areas. There is a need to incorporate malaria control programs alongside CF development. Continuous field monitoring of the incidence of malaria is necessary as the forest condition improves. A continued focus on disease prevention is essential.

NOTES

1. Forestry Administration, "Community Forestry Sub-Decree" (Phnom Penh: Royal Government of Cambodia, 2003).

2. "Independent Forest Sector Review," completed by the Independent Forest Review Team in 2004.

3. Ibid.

4. Rebecca Butterfield, "Community Livelihoods and Common Property Resources in Cambodia," Technical Paper No. 6, Forest Policy Reform Project (Phnom Penh: Associates in Rural Development, Inc. and Department of Forestry and Wildlife, 1998), 24; Ben Vickers and Chris Dickinson, "Report of a National-level Workshop: Heu, Vietnam," 30–31," in *A Fair Share? Experiences in Benefit Sharing from Community-Managed Resources in Asia,* ed. Sango Mahanty, Kenneth Burslem, and Erica Lee (Bangkok, Thailand: RECOFTC, WWF and SNV, 2007), 69–78.

5. Butterfield, "Community Livelihoods," 24.

6. "Summary of Project Achievements, Lessons Learned and Recommendations on the Implementation of the Japan Social Development Fund Grant for Capacity Building for Sustainable Forest and Land Management" (Phnom Penh: FA/RECOFTC, 2010).

7. "Technical Assistance to the Kingdom of Cambodia for Preparing the Tonle Sap Sustainable Livelihoods Project" (Manila: ADB, 2003).

8. "Rural Economic Profile" (Cambodia: SNEC, 2000).

9. "Wet Season Rice and Vegetables: Field Study in Prek Ta Kong Village, Pursat. Water Utilization Program and WUP-FIN Socioeconomic Studies on Tonle Sap 2 - Modeling the Flow Regime and Water Quality of the Tonle Sap" (MRCS/WUP-FIN, Finnish Environment Institute Consultancy Consortium, 2002).

10. See the World Bank's poverty assessment, *Cambodia – Halving Poverty by 2015?* (World Bank, 2006).

11. W. D. Sunderlin et al., *Poverty and Forests: Multi-country Analysis of Spatial Association and Proposed Policy Solutions* (Bogor, CIFOR, 2007).

12. *Towards Wellbeing in Forest Communities: A Source Book for Local Government* (Bogor, Indonesia: CIFOR, 2007), 90.

13. "Technical Assistance to the Kingdom of Cambodia for Preparing the Tonle Sap Sustainable Livelihoods Project" (Manila: ADB, 2003).

14. *An Assessment of Forest Conflicts at the Community Level in Cambodia* (Phnom Penh: USAID, 2004).

15. Ibid., CIFOR, 2007.

16. Sango Mahanty and Michael Nurse, "Introduction to Benefit Sharing in Community-Based Natural Resource Management," in *A Fair Share? Experiences in Benefit Sharing from Community-Managed Resources in Asia,* ed. Sango Mahanty, Kenneth Burslem, and Erica Lee (Bangkok, Thailand: RECOFTC, WWF and SNV, 2007), 3–9.

17. See Kim Sreng Heov, Khlok Bottra, Kasper Hansen and Christian Sloh, "Forest Management Options in Cambodia," *CDRI Policy Brief,* issue 4 (March 2006): 4.

18. Jason Steele, "Cambodia Village Agroforestry: A Pilot Project for Sustainable Energy Resources, Income Generation, and Gender Equity" (VA, USA: Virginia Tech, 2007), 67.

19. Sunderlin et al., *Poverty and Forests.*

20. Butterfield, "Community Livelihoods."

21. Leang Seng, Pouk Bunthet, and Dul Vuth, "Community-based Protected Area Management and Sustainable Livelihoods," in *Cambodia's Community Forestry Research Project: Selected Experiences and Lessons Learned,* ed. Irwin Ken, Sy Ramony, Toby Carson and Ken Serey Rotha (Phnom Penh, Cambodia: IDRC, DNCP/MoE, RUA, CFO-FA-MAFF and CBNRM-LI, 2006), 77–84.

22. Maredy Im, Heng Sokla, and Sok Mary, "Role and Involvement of the Commune Council in Community Forestry Activities in Domnak Neak Ta Thmor Puan," in *Cambodia's Community Forestry Research Project: Selected Experiences and Lessons Learned,* ed. Irwin Ken, Sy Ramony, Toby Carson and Ken Serey Rotha (Phnom Penh, Cambodia: IDRC, DNCP/MoE, RUA, CFO-FA-MAFF and CBNRM-LI, 2006), 51–64.

23. Yam Malla, "Poverty, Environmental Sustainability, Forest Conservation and Governance: What Contributions from Community Forestry?" (Paper Presented at the Regional Conference on Land Administration and Poverty Reduction and Sustainable Development, Hanoi: Vietnam Association of Geodesy, Cartography and Remote Sensing, and Fredskorpset [Norwegian Development Agency], 2006), 14–158.

24. S. Suyanto, Noviana Khususiyah, and Beria Leimona, "Poverty and Environmental Services: Case Study in Way Besai Watershed, Lampung Province," *Indonesia Ecology and Society* 12, no. 2 (2007).

25. Stephen R. Tyler, *Focus: Co management of Natural Resources, Local Learning for Poverty Reduction* (Ottawa, Canada: International Development Research Centre, 2006), 89.

26. Merril Halley, Kong Boravuth, Pet Phaktra, Phay Somany, and Prak Thearith, "Report of a National-Level Workshop, Kratie, Cambodia," in *A Fair Share? Experiences in Benefit Sharing from Community-Managed Resources in Asia,* ed. Sango Mahanty, Kenneth Burslem, and Erica Lee (Bangkok, Thailand: RECOFTC, WWF and SNV, 2007), 13–26.

27. Malla, "Poverty."

28. Karel Mayrand and Marc Paquin, "Payments for Environmental Services: A Survey and Assessment of Current Schemes" (Montreal, Canada: Uniséra International Centre, Commission for Environmental Cooperation of North America, 2004), 60.

29. "Focusing on Cambodia's High Value Forests: Livelihoods and Management" (Wildlife Conservation Society, 2004).

30. Vickers and Dickinson, "Report of a National-level Workshop."

31. N. Landell-Mills, "Marketing Forest Environmental Services – Who Benefits?" (London: International Institute for Environment and Development, Gatekeeper Series, 2002), no. 104, http://www.iied.org/docs/gatekeep/GK104.pdf; N. Landell-Mills and I. Porras, "Silver Bullet or Fools' Gold: A Global Review of Markets for Environmental Services and Their Impact on the Poor" (London: International Institute for Environment and Development, 2002).

32. Steele, "Cambodia Village Agroforestry."

33. "Report of Launching Workshop on Avoided Deforestation Community Forestry Carbon Pilot Project in Oddor Meanchey Province, Cambodia (FA/Community Forestry International, 2008).

34. E. Lee, B. Leimona, M. van Noordwijk, C. Agarwal and S. Mahanty, "Payments for Environmental Services: Introduction to Feasibility, Supplier Characteristics and Poverty Issues," in *Notes from the Field,* ed. Mikaela Rosander, Issue 2 (Bangkok, Thailand: RECOFTC; Bogor, INDONESIA: RUPES Program, The World Agroforestry Centre, Southeast Asia; Gurgaon, India: Winrock International India, 2007).

IMPLEMENTATION OF THE NATIONAL COMMUNITY FORESTRY PROGRAM

MARK POFFENBERGER
LONG RATANAKOMA

THE NCFP IDENTIFIES A NUMBER OF SUBPROGRAMS OF community forestry activities, including the following: (1) community forestry identification and formalization, (2) community, institutional, and livelihood development, and (3) community forestry development support. These initiatives are designed to facilitate the transition to decentralized community-based management of state public forestlands.

COMMUNITY FORESTRY IDENTIFICATION AND FORMALIZATION

The first component of the NCFP is to engage communities in the management of state public forest lands and to improve their management. One of the most important activities in building the national CF program is to identify existing or potential CF groups and to recognize their rights to manage local forests, upon which they depend for their livelihoods and key environmental services. Many forest communities have utilized local forests for generations, but their rights have never been formalized. Further, in many cases forest-dependent communities may be in conflict with concessionaires and other interest groups. Under the NCFP, community forestry sites are formally created, with land use conflicts resolved and forests protected from conversion to alternative uses.[1]

The purpose of the first subprogram or NCFP component is to identify where CF groups exist, clarify their status, resolve any outstanding conflicts, and formally recognize their existence. In addition, it seeks to identify where new CF sites could be developed. Identification of potential areas for community forestry has been essential in developing the NCFP. In addition to indigenous and traditional CF systems that have been operating in Cambodia for generations, new groups have emerged in the past two decades in response to threats to local forests. According to Cambodia's NFP, the NCFP has identified eleven steps in the identification and formalization process.

COMMUNITY, INSTITUTIONAL, AND LIVELIHOOD DEVELOPMENT

The second NCFP component involves supporting community, institutional, and livelihood development in order to benefit forest-dependent households and villages. Under this initiative, attention is directed towards building the capacity of CF management institutions to encourage strong participatory governance through consensus-based decision-making. The development of participatory planning and budgeting systems, forest protection strategies, and conflict management skills are also emphasized. This subprogram is also designed to enhance the livelihoods of forest-based villages by providing training in business management, marketing, and the technical aspects of specific products. Partnerships with private sectors are encouraged, and CF facilitators will seek to arrange equitable contracts, with independent certification required for specific markets.

Screening of CFMCs will be conducted to identify high-potential CF areas for specific institutional and enterprise support activities. Activities may include formal training, as well as informal mentoring. Capital will be secured to cover start-up costs for CF enterprises through grants from donors and loans from financial institutions. Micro-credit institutions will also be developed. At present, subprogram two is in the planning phase and is intended to build upon the CFMC institutions emerging through subprogram one.

COMMUNITY FORESTRY DEVELOPMENT SUPPORT

The third NCFP component supports the further strengthening of CF legislation and policies, as well as the formulation of a national CF information management system, and facilitates coordination between the RGC, the FA, local authorities, and communities. The program will guide a national review of all four CF strategies after five years (2014) to broaden the scope and flexibility of existing programs. One expected result is to make CF policies and programs more enabling, and "to give CF supremacy over competing land use claims and forest management, with long-term secure tenure."[2] The review would also make changes to the Forestry Law that would give commune councils more control over forest resources. The process would also seek to unify under a single framework the four CF modalities: community forestry, community-based production forestry, partnership forestry, and community conservation forestry. The implementation process for engaging, formalizing, planning, and implementing CF is characterized by 11 steps.

Step 1: Establishment
The objective of this initial activity is to determine the interest of the community in managing state public forestlands. Under this step, interested communities submit a formal application, certified by the village chief, to the commune and district councils, as well as to the provincial governor, showing that they have complied with the CF group formation requirements and that at least 60 percent of the community members support the establishment of a CF group. Both FA field staff and NGOs often act as facilitators, providing information to communities interested in forming CF groups and helping them to coordinate with local government. Requests for CF group formation may be initiated by the community itself, or it may be catalyzed by the FA or NGOs working in the area.

Step 2: Information Gathering
Once it is determined that a community is ready to participate in the NCFP, it is important to acquire information regarding the location, stakeholders, village conditions, and forest use practices. During Step 2, a working group is formed with members from the FA, community, commune, and participating NGOs to collect information about the

village. Meetings with commune councils and local communities are combined with awareness-raising concerning the value of forest conservation and the benefits available under the NCFP. Where communities, commune councils, and local Department of Land offices are interested, potential CF sites are integrated into Commune Land Use Plans (CLUPs). Data on existing and potential CF sites are also integrated.

Step 3: Establishment of a CF Management Committee (CFMC)
The next step in the process is to create a CF management committee (CFMC). A temporary committee may be formed to identify candidates and hold an election of officers. Once the officers are identified and the CFMC is formed, a formal request can be sent to the FA cantonment office to establish a CF area. Orders (*Deikas*) are issued by the commune council recognizing the CFMC. This process often moves forward rapidly, reflecting strong community interest in gaining recognized rights to local forests. One initiative, the Capacity Building for Sustainable Forest and Land Management Project (CBSFLMP), had projected eighty CFMCs would be established in its five-province target area between 2006 and 2010; however, due to strong demand, 220 CFMCs were formed during this period.[3] Reports from other provinces suggest grassroots demand for CFMC establishment is also strong, especially in areas with good forests under threat. An important constraint to accelerating the creation of CFMCs is the limited staff and financial resources required by the FA and NGO community to respond to the demand.

Step 4: Preparation of CFMC Bylaws
For the CF governance structure to function effectively, the organization's officers and members must develop bylaws to guide the election of officers, clarify the rights and responsibilities of members, and oversee the operations of the CFMC. The bylaws are developed by the CFMC board of directors and the community members, with assistance from the FA and NGOs as required. Bylaws are approved by the CFMC and the commune council and provided to community members for implementation.

Step 5: Demarcation of CF Boundaries and Mapping
The boundary of each community forest is demarcated and mapped during Step 5, with assistance from the FA and NGOs. GPS devices are used to

generate an accurate outline of the spatial area of each CF site. This data is later downloaded into FA and NGO geographic information systems (GIS) to generate maps of the CF forest area. CF groups are advised to invite representatives from adjacent villages to participate in the forest demarcation process in order to avoid conflicts over boundaries. Concrete boundary markers produced locally, to be consistent with FAO standards, are utilized as permanent posts. The mapping process is an essential step in clarifying the location of community forests and resolving local conflicts. Maps are then included in CF management plans and also shared with the land office, commune council, and district offices to better coordinate with land use planning activities.

Step 6: Preparation of CF Regulations
Rules and regulations are an essential component of any CF management system. During Step 6, CFMC members meet to determine appropriate guidelines for sustainable forest use and conservation. This may include NTFP and timber harvesting levels, special conservation areas, and other forest utilization guidelines. Rules may include fines and sanctions for those who do not follow CFMC regulations. Once the regulations are drafted and approved by the CFMC board and members, they are submitted for recognition to the commune council, district authority, and FA cantonment.

Step 7: Preparation and Approval of CF Agreement
The community forestry agreement is a bilateral contract between the FA and the CFMC to manage the site. It is the result of a process that begins with Step 1 and the establishment of the CFMC. Key benchmarks include approval of the CFMC and CF site location by the local FA office, national FA, and MAFF. Once approval from all three government levels has been secured, along with those from the commune council and district and provincial governments, the CF agreement may be signed by the CFMC board of directors and the FA cantonment or other authority. According to the NFP:

> By completing this process, the community has secured access to the forest resources for an agreed period of time (usually fifteen years with additional fifteen-year renewals), that will

only be revoked if the community fails to meet the requirements outlined in the respective legal documents.[4]

Step 8: Preparation and Approval of CF Management Plan

CF management plans include long-term goals for CF areas, including both production and conservation. These plans may identify timber and non-timber forest production levels, rotation timing, processing activities, and marketing strategies. They may also identify critical watersheds for conservation and restoration, areas of high biodiversity value, ecotourism enterprises, and other activities. They must reflect strategies that propose equitable and sustainable use of the forest resources in ways that benefit the entire community.

The preparation of a CF management plan requires broad-based community participation, training, information gathering, data analysis, and further mapping. It is expected that most CF groups will not initiate the formulation of management plans until they are well-established organizations with effective protection mechanisms in place. Once communities believe they have sufficiently capable community forestry management committees and broad-based community participation in patrolling, fire control, and other protection activities, they may seek FA and NGO assistance in preparing a detailed CF management plan. When the plan is drafted, it may be submitted for review and approval to the FA cantonment office. By 2010, only four CFMCs had submitted management plans for approval, covering 432 hectares. These were all in Siem Reap Province, where the FAO had provided assistance to the local FA and CFMCs to develop sustainable pole harvesting management plans.[5]

Step 9: Enterprise Development

Enterprise development represents an option for some community production forests, as well as more general community forests. Requirements for enterprises will depend on the nature of the activity. Enterprises may involve sustainable timber management, NTFP processing and marketing, ecotourism, and other for-profit initiatives that benefit the participating communities. Honey production and rattan harvesting and processing are two enterprises where CF groups are developing experience.

Step 10: Implementation of the Forest Management Plan

Once community forest management plans have been approved, they will be implemented by participating CFMCs, along with any associated enterprises. Annual work plans will be developed and reviewed by the general assembly of the CFMC, with end-of-the-year reviews to assess progress.

Step 11: Monitoring and Evaluation

Monitoring and evaluation will be carried out in coordination with reviews in accordance with the forest management plan. CFMCs will be responsible for providing information on their forests to the FA concerning forest condition, management activities, and related issues. In some cases, such as REDD+ project areas, more detailed monitoring will be required to track changes in carbon stocks and socioeconomic conditions.

By contrast, CF programs that are "target driven" and implemented without community outreach, participation, and capacity building tend to create CF institutions that may perform poorly in long-term forest management. It is important that the FA and its NGO partners seek to create a pool of potential CF sites where the communities are genuinely interested in conserving and sustainably managing local forests, which is often reflected in early activities, such as outreach, education, training, forest patrols, planning meetings, and control of illegal activities. The NCFP will also need to provide new forest-based economic opportunities to create financial incentives for communities to participate in forest management activities.

SUBMISSION REQUIREMENTS OF THE FA/MAFF

Many CF sites that are submitted to FA/MAFF for approval lack key information and documentation. This creates substantial delays in processing and added costs to the project, while undermining community motivation and delaying management activities. Requirements for documentation must be clear, and submissions need to be supported by local governments and FA field staff. Proper preparation is required to ensure that CF applications are complete and meet the requirements of the FA. It may be cost-effective to organize regular training and awareness-raising workshops at the district or provincial level that include a group of potential CFMCs and local government representatives.

Through closer coordination and clustering sites, the CF proposals can progress through the approval pipeline together. Experiences from Oddar Meanchey Province indicate that processing a cluster of CF applications led to most being approved in a comparatively short period of time. As Table 1 indicates, several inspectorates and provinces have lagged behind others and require additional technical and financial support to accelerate the processing of CF group approvals.

CAPACITY BUILDING OF CFMC

While the series of steps leading to CF Agreements (steps 1–7) and to productive management (steps 8–11) provide a clear pathway to building capacity of CFMCs, these activities should be performed with community participation, education, and training. Without participation and information, CFMCs may progress through the pipeline without developing the necessary skills and, as a consequence, not function fully. CFMC bylaws, rules, and regulations may be poorly understood by many community families.

There is a need to establish benchmarks that indicate a CF group has gained institutional and management capacities before it signs a CF agreement. Clustering and networking CF groups at the sub-district and district levels to evaluate each other's capacities may be helpful in getting peer evaluations and building management partnerships. Past experiences in Cambodia with provincial CF federations indicate that such networks and apex bodies can provide a framework for training and dialogue, which can accelerate the development of CF systems on a landscape level.

SUBGRANTS TO CFMCS, FA FIELD OFFICES, AND LOCAL NGOS

A number of community forestry support programs have provided small grants to CFMCs, FA field offices, and local NGOs to provide them with resources for CF identification, CFMC activities, and CFMC training and capacity building. Due to the absence of funding at the local level to build CF groups, seed and small grant programs provide a low-cost, high-impact strategy to mobilize CF activities. Small grant programs need to be expanded and extended nationally to reach scattered and isolated forest communities and to be integrated with systematic training and capacity building programs.

There is an urgent need in Cambodia to protect and sustainably manage state public forests that are currently threatened by agricultural expansion, mining, illegal logging, and commercial development. Implementation of the NCFP represents a cost-effective strategy that conserves the nation's forest resources while securing community resource tenure rights and addressing the economic needs of the country's rural people. Much of Cambodia's forests could be sustainably and productively managed by communities if adequate technical and financial support is provided. Progress achieved in expanding and formalizing CF groups over the past five years demonstrates the commitment of the RGC, the MAFF, and the FA to move forward with this national program. Support from development agencies will be instrumental in providing funding for institutional development, training, small grants, and field programs.

Over the past decade, it is evident that those regions of Cambodia that received substantial external support for CF development, particularly the North Tonle Sap Inspectorate, have experienced accelerated expansion and formalization of their CF groups. In the North Tonle Sap Inspectorate, CF support began with a joint collaboration between the RGC and the FAO in Siem Reap Province and was expanded to Kompong Thom and Oddar Meanchey through the efforts of local and international NGOs, including CFI, Project Concern, Oxfam G.B., Pact, RECOFTC, and others, all of which worked closely with local offices of the FA. This type of collaboration between communities, local government, the FA, and civil society organizations funded through support from development agencies is required to fully implement the NCFP and achieve national goals for stabilizing forest cover at 60 percent and addressing the livelihood needs of the rural population.

Table 1. Cambodia Community Forestry Statistics

Province	Existing CFs			Approved by MAFF			Signed CFA		
	No. of CFs	Area in ha	No. of villages	No. of CFs	Area in ha	No. of villages	No. of CFs	Area in ha	No. of villages
1. Mekong FA Inspectorate									
Ratanakiri	31	20,699	54	9	908	17			
Kratie	35	54,357	45	11	21,558	15			
K. Cham	10	3,480	10	5	1,783	5			
Mondolkiri	4	4,176	4	-	-	-			
Stung Treng	6	14,838	18	-	-	-			
Svay Rieng	2	504	4	2	504	4			
2. North Tonle Sap Inspectorate									
K. Thom	82	80,908	102	57	48,479	67	52	45499	62
Siem Reap	37	18,122	62	37	18,122	62	35	17,523	60
Preah Vihear	21	45,344	33	18	37,063	28			
Oddar Meanchey	14	65,168	54	14	65,168	54	13	64,318	52
Bateay Meanchey	11	4,970	20	11	4,970	20	11	4,970	20
3. South Tonle Sap Inspectorate									
Pursat	56	7,286	56	38	2,783	38	12	731	12
K. Chhnang	33	10,910	61	15	6,715	39	15	6,715	39
K. Speu	22	12,563	34	15	5,877	20	5	4,188	5
Battambang	27	12,530	55	8	3,815	16			

Pailin	4	858	11	4	858	11			
4. Gulf Inspectorate									
Kampot	27	12,530	55	8	3,815	16	7	3,784	15
Koh Kong	13	15,093	3	13	15,093	3	2	3,638	3
Takeo	13	10,791	61	1	557	10			
Kep	3	1,009	3	1	60	1	1	60	1
Total	441	389,021	710	272	237,844	426	153	151,426	269

Source: FA[6]

NOTES

1. Department of Forestry and Wildlife, Final Draft National Community Forestry Strategic Plan (DOFW, Phnom Penh, 2000), 79.

2. Ministry of Agriculture, Forestry and Fisheries, Cambodia's National Forest Programme, MAFF (Phnom Penh, 2010), 87.

3. RECOFTC 2010. Summary of Project Achievements, Lessons Learned, and Recommendations on the Implementation of the Japan Social Development Fund Grant for Capacity Building for Sustainable Forest and Land Management (Grant TF055390). (Phnom Penh, September 2006 to June 2010), 12.

4. Ministry of Agriculture, Forestry and Fisheries, 82.

5. RECOFTC 2010, 23.

6. Ibid., 32.

BIBLIOGRAPHY

ADB. "Technical Assistance to the Kingdom of Cambodia for Preparing the Tonle Sap Sustainable Livelihoods Project." Manila: ADB, 2003.

AFD-Cambodia. February 2011. *Case Study on the Links between Migrations, Agricultural Expansion, and Deforestation in the Pailin Province.* AFD, Jérémie Dulioust, Mai 2011.

ARD. "Proceedings of Meeting to Brief the International Community and Senior Officials of the Royal Government of Cambodia on Forest Policy Reform." Phnom Penh: ARD, Inc./MAFF,1998.

Alcorn, J. B. "Factors Influencing Botanical Resource Perception among the Huastec: Suggestions for Future Ethnobotanical Inquiry. *Journal of Ethnobiology* 1, no. 2 (1984.): 221–30.

Arthur, R. and R. Friend. "Inland Capture Fisheries in the Mekong and their Place and Potential within Food-led Regional Development." *Global Environmental Change* 21, no. 1 (2010): 219–26.

Arthur, R., R. Friend, and M. Marschke. 2011. "Making Adaptive Co-management more than a Marriage of Convenience: Reconciling Theory and Practice in the Management of Fisheries in the Mekong Region." In *Collaborative Resilience: Moving From Crisis to Opportunity*, ed. B. Goldstein. MIT Press, 2011.

Ashwell, David A. *Cambodia: A National Biodiversity Prospectus.* Phnom Penh: IUCN, 1997.

————. "Cambodia's National System of Protected Areas." In *Cambodia: A National Biodiversity Prospectus.* Phnom Penh: IUCN, 1997. Pp. 60–70.

Asian Development Bank. 2004. *Future Solutions Now: The Tonle Sap Initiative.* Manila.

An Assessment of Forest Conflicts at the Community Level in Cambodia. Phnom Penh: USAID, 2004.

Baird, Ian G. *The Ethnoecology, Land-Use, and Livelihoods of the Brao-Kavet Indigenous Peoples in Kok Lak Commune, Voen Say District, Ratanakiri Province, Northeast Cambodia.* Ban Lung: NTFP, 2000.

————. "The Case of the Brao: Revisiting Physical Borders, Ethnic Identities and Spatial and Social organisation in the Hinterlands of Southern Laos and Northeastern Cambodia." In *Recherches Nouvelles sur le Laos*, ed. Y. Goudineau and M. Lorrillard. Etudes thématiques No. 18, EFEO. Paris and Vientiane, 2008. Pp. 595–620.

————. "Various Forms of Colonialism: The Social and Spatial Reorganisation of the Brao in Southern Laos and Northeastern Cambodia." Ph.D. Dissertation, Geography Department, The University of British Columbia, Vancouver, Canada, 2008.

————. "Controlling the Margins: Nature Conservation and State Power in Northeastern Cambodia." In *Development and Dominion: Indigenous Peoples of Cambodia,*

Vietnam and Laos, ed. Frédéric Bourdier. Bangkok: White Lotus Press, Bangkok, 2009. Pp. 215–48.

──────. *Dipterocarpus Wood Resin Tenure, Management and Trade: Practices of the Brao in Northeast Cambodia.* Saarbrücken, Germany: Verlag Dr. Müller, 2009.

Baird, I. G., and P. Dearden. "Biodiversity Conservation and Resource Tenure Regimes – A Case Study from Northeast Cambodia." *Environmental Management* 32, no. 5 (2003): 541–50.

Baird, I. G., K. Tuptim, and M. Baird. "The Kavet and the Kreung. Observations of Livelihoods and Natural Resources in Two Highlander Villages in the Districts of Veun Say and Ta Veng, Ratanakiri Province, Cambodia." Livelihoods and Natural Resources Study, Novib and Oxfam (UK and Ireland), Ratanakiri, 1996.

Balee, W. and A. Gely. "Managed Forest Concession in Amazonia: The Ka'apor Case." *Advances in Economic Botany* 7 (1989): 129–58.

Barrera-Bassols, N. and V. M. Toledo. "Ethnoecology of the Yucatec Maya: Symbolism, Knowledge and Management of Natural Resources." *Journal of Latin American Geography* 4, no. 1 (2005): 9–41.

Basso, K. H. *Wisdom Sits in Places.* Albuquerque: University of New Mexico Press, 1996.

BBC News – Asia. "Cambodia Suspends New Land Grants for Companies." BBC, May 7, 2012.

Berkes, F. "Traditional Ecological Knowledge in Perspective." In *Traditional Ecological Knowledge: Concepts and Cases,* ed. J. T. Inglis. Ottawa: International Program on Traditional Ecological Knowledge and International Development Research Centre, 1992.

──────. "Devolution of Environment and Resources Governance: Trends and Future." *Environmental Conservation* 37, no. 4 (2010): 489–500.

Bird, J. "A Responsible Approach to Building Dams on the Mekong." *The Nation,* September 22, 2008. Retrieved online 2011: http://chrislang.files.wordpress.com/2008/11/the_nation_22_sept_08.pdf.

BPAMP [Biodiversity and Protected Area Management Project]. "Virachey National Park Management Plan 2003–2007." Ministry of Environment, Phnom Penh, 2003.

Bradley, Amanda. "Communities and Carbon: Establishing a Community Forestry-REDD Project in Cambodia." Phnom Penh: Pact, 2009.

──────. "Does Community Forestry Provide a Suitable Platform for REDD? A Case Study from Oddar Meanchey, Cambodia." In *Lessons about Land Tenure, Forest Governance, and REDD+,* ed. Lisa Naughton-Treves and Cathy Day. Madison, Wisconsin: The Land Tenure Center, 2012.

Bradley, Amanda and Robert Oberndorf. "Buddhism and the Role of the Pagoda in Community Forestry Development in Cambodia." Report by Community Forestry International. Phnom Penh, March 2005.

Bradley, Amanda, Mark Poffenberger, and Ponreay. "National Survey of Community Forestry in Cambodia." Draft Document. Phnom Penh, Cambodia: Community Forestry Alliance for Cambodia, 2005.

Bradley, Amanda and Phuong Pichponreay. "Community Forestry Seed Grants." Phnom Pehn: CFI, 2006.

Bradley, Amanda and Andrew McNaughton. "Community Forestry and Honey Enterprise Development." Phnom Penh: CFI, 2007.

Brinkley, Joel. "Cambodia's Curse." *Foreign Affairs,* March/April 2009, 113.

Brody, H. *Maps and Dreams: Indians and the British Columbia Frontier.* Vancouver/Toronto Douglas & McIntyre, 1981.

Brown, Graeme and Alistair Stephens. *Mapping of Possible Community Forest Management Tenures in Ratanakiri Province, Cambodia.* Phnom Penh, Cambodia: Community Forestry Alliance for Cambodia (CFAC), 2005.

Brown, Graeme, Jeremy Ironside, Mark Poffenberger, and Alistair Stephens. "Formalizing Community Forestry in Ratanakiri Province." In *Cambodia: Linking Indigenous Resource Systems to Government Policies and Programs.* CFI: Phnom Penh, Cambodia, 2007.

———. "Forest Stewardship in Ratanakiri: Linking Communities and Government." Phnom Penh: CFI, 2006.

Butterfield, Rebecca. "Community Livelihoods and Common Property Resources in Cambodia." Technical Paper No. 6, Forest Policy Reform Project. Phnom Penh: Associates in Rural Development, Inc. and Department of Forestry and Wildlife, 1998.

Cambodia from Recovery to Sustained Development. World Bank: Washington DC, 1996.

Cambodia: Forest Policy Assessment. World Bank/FAO/UNDP, 29 April 1996.

Cambodia Timber Industry Association. *The Equitability of the Forest Taxation System in Cambodia.* Phnom Penh: KMPG International Forestry and Environmental Advisory Services Group, 2001.

Cambodia: Forest Concession Management and Control Pilot Project. Washington, D.C.: World Bank Inspection Panel, 2006.

Carle, Jim. "Reforestation and Natural Forest Rehabilitation Policy in Cambodia." *Forest Policy Reform Project.* Technical Paper No. 3. 1998.

Carson, Toby and Hou Kalyan. "Overview of the Past and Present Situation of CBNRM in Cambodia." In *Emerging Trends, Challenges and Innovations for CBNRM in Cambodia.* Phnom Penh: CBNRM Learning Instute, 2009.

Casagrande, D. G. "Conceptions of Primary Forest in a Tzeltal Maya Community: Implications for Conservation." *Human Organization* 63, no. 2 (2004): 189–202.

CBNRM Learning Institute. *CBNRM in Cambodia: Selected Papers on Concepts and Experiences.* Vol. 1. Phnom Penh: CBNRM-LI, 2005.

———. *Emerging Trends, Challenges and Innovations for CBNRM in Cambodia.* Vol. 2. Phnom Penh: CBNRM Learning Institute, 2009.

CFI eds. "Proceedings of the Non-Timber Forest Product (NTFP) Workshop and Seminar." Phnom Penh: CFI, 2006.

CFI. "Factsheet: Community Forestry Carbon Offset Project." Phnom Penh: Community Forestry International, June 2008.

Chhatre, Ashwini and Arun Agrawal. "Trade-offs and Synergies Between Carbon Storage and Livelihood Benefits from Cabon Commons." *PNAS* 106, no. 42 (2009): 17667–70.

Community-based Forest Protection: An Option for Cambodia. Phnom Penh: NGO Forum, 2007.

Conklin, H. C. 1957. *Hanunoo Agriculture: A Report on an Integral System of Shifting Cultivation in the Philippines.* Forestry Development Paper no. 12. Food and Agriculture Organization of the United Nations, Rome.

"Consolidated ESIA Report as integral part of the Strategic Forest Management Plan" for Kratie/Strung Treng Forest Concession of Everbright CIG Wood Co. Ltd., 2003.

"Cry From The Forest: A 'Buddhism and Ecology' Community Learning Tool." Phnom Penh: The Buddhist Institute in cooperation with the NGO Working Group for Non-formal Monk Environmental Education Project (MEEP) and with support of UNDP-ETAP in collaboration with UNESCO, 1999.

Delang, C. O. "Deforestation in Northern Thailand: The Result of Hmong Farming Practices of Thai Development Strategies?" *Society and Natural Resources* 15 (2002):
483–501.

Denevan, W. 1984. "Ecological Heterogeneity and Horizontal Zonation of Agriculture in
the Amazon Floodplain." In *Frontier Expansion in Amazonia*, ed. M. Schmink and C.
H. Wood. Gainesville: University of Florida Press. Pp. 311–36.

Department of Forestry and Wildlife. "Final Draft National Community Forestry Strategic
Plan." Phnom Penh: DoFW, 2000.

Devichand, Mukul. "Has Cambodia Become a Country for Sale?" BBC News – Asia Pacific,
January 12, 2011.

Dunn, F. L. *Rain-Forest Collectors and Traders: A Study of Resource Utilization in Modern and
Ancient Malaya.* Monographs of the Malaysian Branch, Royal Asiatic Society No. 5.
Kuala Lumpur, 1975.

Economic Land Concessions in Cambodia: A Human Rights Perspective. Phnom Penh:
UN-HCHR, 2007.

Emerson, B. "The Natural Resources and Livelihood Study, Ratanakiri Province, NE
Cambodia." The Non-Timber Forest Products (NTFP) Project, Ratanakiri Province,
1997.

Evans, Tom D., Hout Piseth, Phet Phaktra, and Hang Mary. *A Study of Resin Tapping and
Livelihoods in Southern Mondulkiri, Cambodia with Implications for Conservation
and Forest Management.* Phnom Penh: World Conservation Society, 2003.

Evans, P., M. Marschke, and K. Paudyal. *Flood Forests, Fish and Fishing Villages, Tonle Sap,
Cambodia: Community Forest Management Trends in Southeast Asia.* Research
Network Report No. 12. Asia Forest Network, 2004.

FAO. *The State of the World Fisheries and Aquaculture 2010.* Rome: Food and Agricultural
Organisation of the United Nations, 2010.

Fichtenau, Jurgen, Ly Chou Beang, Nup Sothea, and Dy Sophy. "Assessment of Ongoing
Community Forestry Initiatives in Cambodia." Phnom Penh: DFoW/GTZ, 2002.

"Findings and Recommendations of the Log Monitoring and Log Control Project: Main
Report." Submitted to the Ministry of Agriculture, Forestry and fisheries, Royal
Government of Cambodia. Bethesda, MD: DAI, 1998.

Forest Policy Transition Paper for Cambodia. Phnom Penh: Associates for Rural Development, Inc., 1998.

Forestry Administration. "Community Forestry Sub-Decree." Phnom Penh: Royal Government of Cambodia, 2003.

Fox, Jefferson, Dennis McMahon, Mark Poffenberger, and John Vogler. *Land for My
Grandchildren.* Cambodia: CFI/EWC, 2008.

Fraser, Thomas, GFA-AGRAR, ANZDEC. "Cambodia Forest Concession Review Report."
Sustainable Forest Management Project, Asian Development Bank (TA-3152-Cam),
Ministry of Agriculture, Forestry and Fisheries, General Directorate of Forestry,
Royal Government of Cambodia, Phnom Penh, 2000.

GIZ. "Land Distribution in Cambodia – Experiences and New Approaches for State Land
Management." FIG Working Week 2011, Marrakech, Morocco, 18–22, May 2011.
http://www.fig.net/pub/fig2011/ppt/ts04b/ts04b_mueller_5376_ppt.pdf

Global Witness. *Deforestation without Limits.* London: Global Witness, 2004.

Goudineau, Y., ed. " Resettlement and Social Characteristics of New Villages: Basic Needs
for Resettled Communities in the Lao PDR." Main Report. Volume 1. Ostrom,
UNESCO and UNDP, Vientiane, 1997.

———. "Resettlement and Social Characteristics of New Villages: Basic Needs for Resettled Communities in the Lao PDR." Provincial Reports. Volume 2. Ostrom, UNESCO and UNDP, Vientiane, 1997.

Haggan, N., B. Neis, and I. G. Baird, eds. *Fishers' Knowledge in Fisheries Science and Management*. Coastal Management Sourcebooks Series, UNESCO, Paris, 2007.

Halley, Merril, Kong Boravuth, Pet Phaktra, Phay Somany, and Prak Thearith. "Report of a National-Level Workshop, Kratie, Cambodia." In *A Fair Share? Experiences in Benefit Sharing from Community-Managed Resources in Asia,* ed. Sango Mahanty, Kenneth Burslem, and Erica Lee. Bangkok, Thailand: RECOFTC, WWF and SNV, 2007. Pp. 13–26.

Harris, Ian. *Cambodian Buddhism: History and Practice.* Bangkok: University of Hawaii Press, 2005.

Heov, Kim Sreng, Khlok Bottra, Kasper Hansen, and Christian Sloh. "Forest Management Options in Cambodia." CDRI Policy Brief. March 2006, issue 04.

Huljus, Hohannes, and Britta Jell, *Cambodia: Review of Strategic Forest Management Plans Prepared by Concession Companies Operating in Cambodia; Part II.* Hamburg, Germany: GFA Consulting Group, 2005.

Hunn, E. S. *Nch'I-Wana "The Big River": Mid-Columbia Indians and their Land.* Seattle: University of Washington Press, 1990.

Im, Maredy, Heng Sokla, and Sok Mary. "Role and Involvement of the Commune Council in Community Forestry Activities in Domnak Neak Ta Thmor Puan." In *Cambodia's Community Forestry Research Project: Selected Experiences and Lessons Learned,* ed. Irwin Ken, Sy Ramony, Toby Carson, and Ken Serey Rotha. Phnom Penh, Cambodia: IDRC, DNCP/MoE, RUA, CFO-FA-MAFF and CBNRM-LI, 2006. Pp. 51–64.

Independent Forest Sector Review. *The Forest Sector in Cambodia: Part I – Policy Choices, Issues and Options.* April 2004.

Ingram, G. B. The Need for Knowledge from Indigenous Communities in Planning Networks of Protected Habitat for the Conservation of Biological Diversity: Three Island Settings. In *Ethnobiology: Implications and Applications. Proceedings of the First International Congress of Ethnobiology,* ed. D. A. Posey. Belam, Brazil, 1990.

Ironside, Jeremy and Ian G. Baird. "Wilderness and Cultural Landscape: Settlement, Agriculture, and Land and Resource Tenure adjacent to Virachey National Park, Northeast Cambodia." Cambodia: Biodiversity and Protected Area Management Project – BPAMP, DNCP/MOE, 2003.

Johannes, R. E. "Integrating Traditional Ecological Knowledge and Management with Environmental Impact Assessment." In *Traditional Ecological Knowledge: Concepts and Cases,* ed. J. T. Inglis. Ottawa: International Program on Traditional Ecological Knowledge and International Development Research Centre, 1992. Pages 33–39.

Keller, C., J. Jordi, K. Gregerson, and I. G. Baird. "The Brao Dialects of Cambodia: Lexical and Phonological Variations." *Revue de l'Institut de la Langue Nationale de l'Académie Royale du Cambodge.* Institute of Language, Phnom Penh. Special Issue (July 2008): 87–152.

Kent, Alexandra and David Chandler, eds. *People of Virtue: Reconfiguring Religion, Power and Moral Order in Cambodia Today.* Studies in Asian Topics, no. 43. Copenhagen: NIAS Press, 2008.

Keown, Damien. 2005. *Buddhist Ethics: A Very Short Introduction.* Oxford: Oxford University Press.

Korten, David C. "Micro-Policy Reform-the Role of Private Voluntary Development Agencies." In *Community Management: Asian Experiences and Perspectives.* West Hartford, Conn: Kumarian Press, 1987.

Korten, David. *Community Management: Asian Experiences and Perspectives.* New Haven, Conn: Kumarian Press, 1987.

Koy, S. "A Case Study of Virachey National Park, Cambodia." In *Indigenous Peoples and Protected Areas in South and Southeast Asia: From Principles to Practice,* ed. M. Colchester and C. Erni. Copenhagen: International Work Group for Indigenous Affairs, 1999. Pages 134–51.

Landell-Mills, N. "Marketing Forest Environmental Services – Who Benefits?" London: International Institute for Environment and Development, Gatekeeper Series, 2002, no. 104, http://www.iied.org/docs/gatekeep/GK104.pdf.

Landell-Mills, N. and I. Porras. "Silver Bullet or Fools' Gold: A Global Review of Markets for Environmental Services and Their Impact on the Poor." London: International Institute for Environment and Development, 2002.

Lebel, L. "Multi-level Scenarios for Exploring Alternative Future for Upper Tributary Watersheds in Mainland Southeast Asia." *Mountain Research and Development* 26, no. 3 (2006): 263–73.

Ledgerwood, Judy. "Buddhist Practice in Rural Kandal Province 1960 and 2003." In *People of Virtue: Reconfiguring Religion, Power and Moral Order in Cambodia Today,* ed. Alexandra Kent and David Chandler. Studies in Asian Topics, no. 43. Copenhagen: NIAS Press, 2008.

Lee, E., B. Leimona, M. van Noordwijk, C. Agarwal, and S. Mahanty. "Payments for Environmental Services: Introduction to Feasibility, Supplier Characteristics and Poverty Issues." In *Notes from the Field,* ed. Mikaela Rosander. Bangkok, Thailand: RECOFTC; Bogor, INDONESIA: RUPES Program, The World Agroforestry Centre, Southeast Asia; Gurgaon, India: Winrock International India, 2007.

Legerwood, J. and J. Vijghen. "Decision-making in Rural Khmer villages." In *Cambodia Emerges from the Past: Eight Essays,* ed. J. Legerwood. DeKalb, IL: Center for Southeast Asian Studies, Northern Illinois University, 2002. Pp. 176–223.

Letter from the Council of Ministers, Sar. Chor. Nor. No. 699, Council of Ministers, Royal Government of Cambodia, Phnom Penh. May 2008.

Li, T. M. *The Will to Improve: Governmentality, Development, and the Practice of Politics.* Princeton, New Jersey: Duke University Press, 2007.

Loc, V., S. Bush, L. Sinh, and N. Khiem. "High and Low Value Fish Chains in the Mekong Delta: Challenges for Livelihoods and Governance." *Environment, Development, and Sustainability* 12, no. 6 (2009): 889–908.

MacAndrews, Collin. "Strengthening Institutions for Implementation of Forest Policy in Cambodia." Phnom Penh: ARD, Inc. 1998.

MAFF. "Cambodia's National Forest Programme." Phnom Penh: Ministry of Agriculture, Forestry and Fisheries, 2010.

Mahanty, Sango and Michael Nurse. "Introduction to Benefit Sharing in Community-Based Natural Resource Management." In *A Fair Share? Experiences in Benefit Sharing from Community-Managed Resources in Asia,* ed. Sango Mahanty, Kenneth Burslem, and Erica Lee. Bangkok, Thailand: RECOFTC, WWF and SNV, 2007.

Malla, Yam. "Poverty, Environmental Sustainability, Forest Conservation and Governance: What Contributions from Community Forestry?" Paper Presented at the Regional Conference on Land Administration and Poverty Reduction and Sustainable Development. Hanoi: Vietnam Association of Geodesy, Cartography and Remote Sensing, and Fredskorpset (Norwegian Development Agency), 2006.

Marschke, M. "Livelihood in Context: Learning with Cambodian Fishers." Ph.D. Thesis. University of Manitoba, Winnipeg, 2005.

──────. "Le Secteur des Pêcheries de l'Ère « Post » au Cambodge. Une explication de la non-transformation. " *Anthropologie et Sociétés* 32, nos. 1–2 (2008): 133–54.

──────. Forthcoming. "Resource Governance at the Margins: Fish, Trees and Life in Coastal Cambodia. " Ottawa, Canada: University of Ottawa Press.

Marschke, M. and F. Berkes. "Local Level Sustainability Planning for Livelihoods: A Cambodian Experience." *The International Journal of Sustainable Development and World Ecology*, 12, no. 1 (2005): 21–33.

──────. "Exploring Strategies that Build Livelihood Resilience: A Case from Cambodia." *Ecology and Society* 11, no. 1 (2006): 42. [online] URL: http:www.ecologyandsociety.org/vol11/iss1/art42.

Marschke, M. and J. Sinclair. 2009. "Learning for Sustainability through Participatory Resource Management." *Journal of Environmental Management*, 90(1): 206–16.

Maredy Im, Heng Sokla, and Sok Mary. "Role and Involvement of the Commune Council in Community Forestry Activities in Domnak Neak Ta Thmor Puan." In *Cambodia's Community Forestry Research Project: Selected Experiences and Lessons Learned,* ed. Irwin Ken, Sy Ramony, Toby Carson, and Ken Serey Rotha. Phnom Penh, Cambodia: IDRC, DNCP/MoE, RUA, CFO-FA-MAFF and CBNRM-LI, 2006. Pp. 51–64.

Martin, G. J. *Ethnobotany: A Methods Manual.* New York: Chapman & Hall, 1995.

Mayrand, Karel and Marc Paquin. "Payments for Environmental Services: A Survey and Assessment of Current Schemes." Montreal, Canada: Unaiféra International Centre, Commission for Environmental Cooperation of North America, 2004.

McKenney, Bruce. "Questionable Sustainable Concession Forestry in Cambodia." *Cambodia Development Review* no. 6, issue 1 (January–March 2002): 1.

McKenney, Bruce, Yim Chear, Prom Tola, and T. Evans. "Focusing on Cambodia's High Value Forests: Livelihood and Management." Phnom Penh: Cambodia Development Resource Institute and the Wildlife Conservation Society, 2004.

McMahon, Dennis. "Assessment of Community Forestry Sites and Migration Patterns in the Oddar Mean Chey Province, Cambodia." Study conducted for Community Forestry International. Phnom Penh, March 16, 2008.

Meidinger, D. and J. Pojar, eds. *Ecosystems of British Columbia.* British Columbia Ministry of Forestry, Victoria, 1991.

Meta, Prom and Jeremy Ironside. "Effective Maps for Planning Sustainable Land Use and Livelihood." In *Mapping Communities: Ethnics, Values, Practice,* ed. Jefferson Fox, Krisnawati Suryata, and Peter Hershock. East West Center: Honolulu, Hawaii: East-West Center, 2005. Pp. 29–40.

Meyer, C. "Les Nouvelles Provinces: Ratanakiri – Mondulkiri." *Revue Monde en Developement* 28 (1979): 682–90.

Ministry of Agriculture, Forestry and Fisheries. Cambodia's National Forest Programme. MAFF. Phnom Penh, 2010.

Mittelman, A. "Secondary Forests in the Lower Mekong Subregion: An Overview of their Extent, Roles and Importance." *Journal of Tropical Forest Science* 13, no. 4 (2001): 671.

Moran, E. F. 1982. "Limitations and Advances in Ecosystems Research." In *The Ecosystem Concept in Anthropology,* ed. E. F. Moran. AAAS Selected Symposium 92. Boulder, Colorado: Westview Press, Inc. Pp. 3–32.

MRCS. "Wet Season Rice and Vegetables: Field Study in Prek Ta Kong Village, Pursat." Water Utilization Program and WUP-FIN Socioeconomic Studies on Tonle Sap 2

- Modeling the Flow Regime and Water Quality of the Tonle Sap. MRCS/WUP-FIN, Finnish Environment Institute Consultancy Consortium, 2002.

Nasuchon, N., and A. Charles. "Community Involvement in Fisheries Management: Experiences in the Gulf of Thailand Countries." *Marine Policy* 34 (2010): 163–69.

Nathan, Iben, Tove E. Boon, Sovatha Ann, and S. Vanny. "Constraints and Options in Local Forest Management in Cambodia, Is Decentralization a Solution?" Unpublished manuscript, 2009.

Neth, Top, Nobuya Mizoue, Shigetaka Kai, and Toshio Nakao. "Variation in Woodfuel Consumption Patterns in Response to Forest Availability in Kampong Thom Province." *Biomass and Bioenergy* 27 (2004): 61.

Neumann, R. P. *Imposing Wilderness: Struggles over Livelihood and Nature Preservation in Africa.* Berkeley, Los Angeles and London: University of California Press, 1998.

NIS. "General Population Census of Cambodia 1998, Final Result." Phnom Penh: National Institute of Statistics, Ministry of Planning, Phnom Penh, Cambodia, 1998.

NSDP. *National Strategy Development Plan, 2009–2013.* Phnom Penh, Cambodia: Fisheries Administration, 2010.

Nuorteva, P., M. Keskinen, and O. Varis. "Water, Livelihoods and Climate Change Adaptation in the Tonle Sap Lake Area, Cambodia: Learning from the Past to Understand the Future." *Journal of Water and Climate Change* 1, no. 1 (2010): 87–101.

Osborne, Milton. *The Mekong.* New York: Grove Press, 2000.

Oum, P. and C. Ros. "Current Status of the Virachey National Park, Ratanakiri Province." Department of Nature Conservation and Protection, Ministry of Environment, Phnom Penh, 1995.

Oxfam G. B. *Access to Forest Resources and Landlessness: Case Studies of Degraded Forests and Livelihoods in Kampong Thom and Kampong Chhnang Provinces.* Oxfam GB Cambodia: Land Study Project Mini Case Study, 2002.

Peluso, N.L. Coercing conservation? *Global Environmental Change* 3, no. 2 (1993): 199–217.

Poffenberger, Mark. *Keepers of the Forest.* New Haven, Conn.: Kumarian Press, 1990.

Posey, D. A. "Indigenous Ecological Knowledge and Development of the Amazon." *The Dilemma of Amazonian Development,* ed. E. F. Moran. Boulder, Colorado: Westview Press, 1982. Pp. 225–57.

Primack, R. B. *Essentials of Conservation Biology.* 2d ed. Sunderland, Massachusetts: Sinauer Associates, 1998.

Provincial Department of Planning. "Ratanakiri Provincial Development Plan 2006–2010." Provincial Department of Planning, Ban Lung, Ratanakiri, 2005.

Rao, K. and C. Geisler. The Social Consequences of Protected Areas Development for Resident Populations. *Society and Natural Resources* 3, no. 1 (1990): 19–32.

RECOFTC. *Balancing Ownership: Overview of Writing Retreat about the Community Forestry Consultation Process in Cambodia.* Phuket, Thailand: RECOFTC Training and Report Workshop Series, 2002.

———. "Summary of Project Achievements, Lessons Learned, and Recommendations on the Implementation of the Japan Social Development Fund Grant for Capacity Building for Sustainable Forest and Land Management." Phnom Penh: September 2006 to June 2010.

"Report of Launching Workshop on Avoided Deforestation Community Forestry Carbon Pilot Project in Oddor Meanchey Province, Cambodia." FA/Commnity Forestry International, 2008.

RGC. *National Community Forestry Strategic Plan.* Phnom Penh: Department of Forests and Wildlife, 2000.

RGC. "Sar. Chor. Nor. No. 699." Council of Ministers, Kingdom of Cambodia, Phnom Penh, May 2008.

Royal Government of Cambodia. *Final Draft National Community Forestry Strategic Plan.* Phnom Penh: Department of Forests and Wildlife, 2000.

Sahni, Pragati. *Environmental Ethics in Buddhism: A Virtues Approach.* New York: Routledge, 2008.

Sallenave, J. "Giving Traditional Ecological Knowledge Its Rightful Place in Environmental Impact Assessment." *Northern Perspectives* 22, no. 1 (1994): 16–19.

Seng, Leang, Pouk Bunthet, and Dul Vuth. "Community-based Protected Area Management and Sustainable Livelihoods." In *Cambodia's Community Forestry Research Project: Selected Experiences and Lessons Learned,* ed. Irwin Ken, Sy Ramony, Toby Carson, and Ken Serey Rotha. Phnom Penh, Cambodia: IDRC, DNCP/MoE, RUA, CFO-FA-MAFF and CBNRM-LI, 2006.

SNEC. "Rural Economic Profile." Cambodia: SNEC, 2000.

Sneddon, C. "Nature's Materiality and the Circuitous Paths of Accumulation: Dispossession of Freshwater Fisheries in Cambodia." *Antipode* 39, no. 1 (2007): 167–93.

Steele, Jason. "Cambodia Village Agroforestry: A Pilot Project for Sustainable Energy Resources, Income Generation, and Gender Equity." VA, USA: Virginia Tech, 2007.

Steinmetz, R. 1999. "Participatory Biodiversity Surveying and Monitoring in Lao PDR: A Starting Point for Collaborative Management of Protected Areas." Bangkok: World Wide Fund for Nature/Thailand Programme Office, 1999.

———. "The Ecological Science of the Karen in Thung Yai Naresuan Wildlife Sanctuary, Western Thailand." In *Indigenous Peoples and Protected Areas in South and Southeast Asia: From Principles to Practice,* ed. M. Colchester and C. Erni. Copenhagen: International Work Group for Indigenous Affairs, 1999. Pages 84–107.

Steinmetz, R. and R. Mather. "Impact of Karen Villages on the Fauna of Thung Yai Naresuan Wildlife Sanctuary: A Participatory Research Project." *Natural History Bulletin of the Siam Society* 44 (1996): 23–40.

"Summary of Project Achievements, Lessons Learned and Recommendations on the Implementation of the Japan Social Development Fund Grant for Capacity Building for Sustainable Forest and Land Management." Phnom Penh: FA/RECOFTC, 2010.

Sun Tra, Hung and Arvid Sloth. "Review of International Forest Policy Development and Cambodia's Role, Involvement and Potential Benefit." Phnom Penh: Forestry Administration, April 2007.

Sunderlin, W.D. et al. *Poverty and Forests: Multi-country Analysis of Spatial Association and Proposed Policy Solutions.* Bogor, CIFOR, 2007.

Suyanto, S., Noviana Khususiyah, and Beria Leimona. "Poverty and Environmental Services: Case Study in Way Besai Watershed, Lampung Province." *Indonesia Ecology and Society* 12, no. 2 (2007).

Swearer, Donald. "Buddhism and Ecology: Challenge and Promise." Harvard University, <http://fore.research.yale.edu/religion/buddhism/index.html> [accessed July 25, 2009].

Taylor, Jim. "Thamma-Chaat: Activist Monks and Competing Discourses of Nature and Nation in Northeastern Thailand." *Seeing the Forest for the Tree,* ed. Philip Hirsch. Chiang Mai: Silkworm Books, 1996.

Thackway, R. and I. Cresswell. "Toward a Systematic Approach for Identifying Gaps in the Australian System of Protected Areas. In *Ecosystem Monitoring and Protected Areas,* ed. T. B. Herman, S. Bondrup-Nielsen, J.H.M. Willison and N.W.P. Munro. Wolfville, Nova Scotia, Canada: Science and Management of Protected Areas Association, 1995. Pages 473–83.

Towards Wellbeing in Forest Communities: A Source Book for Local Government. Bogor, Indonesia: CIFOR, 2007.

Trandem, A. "A Vietnamese/Cambodian Transboundary Dialogue: Impacts of Dams on the Se San River." *Development* 51 (2008): 108–13.

Tyler, Stephen R. *Focus: Co management of Natural Resources, Local Learning for Poverty Reduction.* Ottawa, Canada: International Development Research Centre, 2006.

United Nations, Cambodia Office of the High Commissioner for Human Rights. "Economic Land Concessions in Cambodia: A Human Rights Perspective." Phnom Penh: United Nations, 2007.

US State Department information on Cambodia online on <http://www.state.gov/r/pa/ ei/bgn/2732.htm\> [accessed June 3, 2009].

Vandergeest, P. "Property Rights in Protected Areas: Obstacles to Community Involvement as a Solution in Thailand." *Environment Conservation* 23, no. 3 (1996): 259–68.

Vickers, Ben and Chris Dickinson. "Report of a National-level Workshop: Heu, Vietnam." In *A Fair Share? Experiences in Benefit Sharing from Community-Managed Resources in Asia,* ed. Sango Mahanty, Kenneth Burslem, and Erica Lee. Bangkok, Thailand: RECOFTC, WWF and SNV, 2007. Pp. 69–78.

Vrieze, Paul and Kuch Narem. "Carving Up Cambodia One Concession at a Time." *The Cambodia Daily,* March 10–11, 2012.

Walker, Robert. "Cambodia: A Land Up for Sale?" BBC online article, on <http://news.bbc. co.uk/1/hi/world/asia-pacific/8144130.stm> [accessed May 18, 2009].

Warner, K. *Shifting Cultivators: Local Technical Knowledge and Natural Resource Management in the Humid Tropics.* Forests, Trees and People, Community Forestry Note no. 8. Food and Agriculture Organization of the United Nations, Rome, 1991.

WCS. "Focusing on Cambodia's High Value Forests: Livelihoods and Management." Wildlife Conservation Society, 2004.

Wester, L. and S. Yongvanit. "Biological Diversity and Community Lore in Northeastern Thailand." *Journal of Ethnobiology* 15, no. 1 (1995): 71–87.

World Bank. *Cambodia at the Crossroads: Strengthening Accountability to Reduce Poverty.* Report No. 30636-KH. A World Bank Document, 2004.

———. *Cambodia-Poverty Assessment 2006.* Phnom Penh: World Bank, 2006.

———. *Cambodia: Halving Poverty by 2015.* Phnom Penh: World Bank, 2006.

———. *Poverty Profile and Trends in Cambodia, 2007.* Findings from the Cambodia Socio-Economic Survey. Poverty Reduction and Economic Management Sector Unit, East Asia and Pacific Region, 2009.

World Bank's Poverty Assessment. *Cambodia—Halving Poverty by 2015?* World Bank, 2006.

Wright, R. G. and J. M. Scott. "Evaluating the Ecological Suitability of Lands for Parks and Protected Areas Using Gap Analysis Databasesin." *National Parks and Protected Areas: Their Role in Environmental Protection,* ed. R. G. Wright. London: Blackwell Sciences, 1995. Pages 121–30.

ABOUT THE AUTHORS

ZAINAL ARIFFIN AHMAD is head of the Graduate Business School, University Tenaga Nasional, Malaysia. He serves as a grand evaluator for the Ministry of Science, Technology and Innovation and the Ministry of Higher Education as well as associate fellow of the Academy of Sciences Malaysia (ASM). He was awarded the National Academy Award for Teaching 2008 and presented on innovations in teaching-learning around the country. He is a member of the Academy of Management (USA).

ABU HASSAN AHMAD is dean and professor of School of Biological Sciences of Universiti Sains Malaysia (USM). His field of specialization is Mosquito ecology and control, agriculture and urban pest control and aquatic entomology. He is a mosquito man of USM who has researched and published many articles and chapters related to mosquitoes. He has successfully supervised over 20 Ph.D. graduates, as well as an external examiner for a number of Malaysian and foreigner universities. He is also attached to several scientific associations, including Malaysian Scientific Association, American Mosquito Control Association, and Chartered Institute of Environmental Health (MCIEH), UK.

IAN G. BAIRD, PH.D. is an assistant professor in the Department of Geography, University of Wisconsin-Madison, USA. He is a native English-speaking Canadian who also speaks Brao, Lao, Thai, and some Khmer. He has been conducting research with the ethnic Brao people of northeastern Cambodia since 1995, and the Brao were the focus of both his M.A. (University of Victoria, Canada) and Ph.D. (University of British Columbia, Canada) research. He has written a number of articles and book chapters about the Brao from both southern Laos and northeastern Cambodia. His two most recent books are titled, *Dipterocarpus Wood Resin*

Tenure, Management and Trade: Practices of the Brao in Northeast Cambodia (Verlag Dr. Müller, 2009) and (with Bruce Shoemaker) *People, Livelihoods and Development in the Xekong River Basin of Laos* (White Lotus Press, 2008).

CHANTAL ELKIN is manager of the Religious Forest Sites Programme at the Alliance of Religions and Conservation (ARC), based in the UK. She worked with Conservation International (CI) for eight years as the Washington-based manager of the Indo-Burma Program and then as Director of CI's Wildlife Trade Program, focusing on the illegal trade in Asia. Chantal holds two master's degrees from the University of London, the first in environment and development in Southeast Asia and the second looking at Buddhism and conservation. Prior to working with CI, Chantal coauthored the 1998 publication *Logging Burma's Frontier Forests: Resources and the Regime* for the World Resources Institute.

JEFFERSON M. FOX, PH.D. is a senior fellow at the East-West Center in Honolulu. He received his doctorate in development studies from the University of Wisconsin-Madison in 1983. He studies land use and land cover change in Asia and the possible cumulative impact of these changes on the region and the global environment. Dr. Fox has coedited several books, most recently *People and the Environment: Approaches for Linking Household and Community Surveys to Remote Sensing and GIS* (Kluwer Academic Press, 2003). His ongoing research includes "Coupled Natural-Human Systems and Emerging Infectious Diseases: Anthropogenic Environmental Change and Avian Influenza in Vietnam," funded by the US National Science Foundation, and "The Expansion of Rubber and Its Implications for Water and Carbon Dynamics in Montane Mainland Southeast Asia," funded by NASA. He has worked in Bangladesh, Cambodia, China (Yunnan), Indonesia, Laos, Nepal, Thailand, and Vietnam.

EDWARD MANINGO is a registered forester with M.S. Degree on Environmental Science and Ph.D. on Technology Management. He was a Technical Advisor of a Community Forestry Project of the Food and Agriculture Organization (FAO) of the United Nations in Siem Reap,

Cambodia in 2008. He was also a associate professor, college dean and research coordinator of the College of Agriculture and Forestry of Negros Oriental State University (NORSU) in Dumaguete City, Philippines in 1997–2012. He with the Department of Environment and Natural Resources (DENR) in the Philippines in 1986–1995 as project manager of a Watershed and Reforestation Project and other positions before joining the Academe.

MELISSA MARSCHKE, PH.D. is an associate professor at the School of International Development and Global Studies, University of Ottawa. Her training is in human-environment relations, with a particular emphasis on livelihoods, common pool resources, and political ecology. She has been researching fisheries and resource governance issues in Cambodia since 1998. She is the author of *Life, Fish and Mangroves: Resource Governance in Coastal Cambodia* (University of Ottawa Press, 2012), and has published in various journals including *Ecology and Society, Global Environmental Change, International Journal of the Commons,* and *Marine Policy.* For more information see http://melissamarschke.wordpress.com.

ROBERT B. OBERNDORF, J.D. graduated from the University of Colorado with a B.A. in Communication in 1989, and from Case Western Reserve University School of Law with a juris doctorate degree in 1993. He is currently admitted to the Ohio Bar (USA) and has been practicing law for almost twenty years. Robert practiced environmental, construction, contract, and real estate law, including litigation, for a regional water pollution control district in the United States for eight years before moving to Asia to work as an international legal consultant in 2001. He has been living in the region since that time, and has worked for the Asian Development Bank, Danida, United Nations Food and Agriculture Organization, International Fund for Agricultural Development, GIZ, East/West Management, Community Forestry International, the Regional Community Forestry Training Center for Asia and the Pacific, World Bank, Wildlife Conservation Society, WWF, International Union for Conservation of Nature, and others on issues relating to natural resources and environmental law, land management and tenure security, climate change/REDD+, renewable energy, governance reform, and international market reform. Robert has experience working in Afghanistan, Brunei, Cambodia, China, East Timor, Indonesia, Lao PDR, Malaysia, the Philippines, Tajikistan, Thailand, United States, and Vietnam.

MARK POFFENBERGER, PH.D. is the executive director of Community Forestry International and a senior fellow at the East-West Center. He has been involved with community natural resource management research, project development, and policy formulation in South and Southeast Asia for thirty-five years. He has informed the development of national community forestry programs in India, Cambodia, the Philippines, and Indonesia, and has written widely on these historic forest sector reforms. He also formed and directed the Asia Forest Network and the Working Group on Community Involvement in Forest Management and contributed to the United Nations' Forum on Forests. Recently, he has been involved in the research and design of a new generation of community-based REDD projects in Cambodia and India. He has written and edited seven books, as well as numerous monographs, articles, and other publications.

LONG RATANAKOMA is the deputy director of the Department of Forest and Community Forestry in charge of community forestry for the Forestry Administration, MAFF, in Cambodia. He has a master's degree in development management and a bachelor's degree in forest science. He facilitates and coordinates with key stakeholders, local Forestry Administration, local authorities, NGO partners, and local community to establish and legalize community forestry in Cambodia. He helped develop the National Forest Program, particularly the National Community Forestry Program, which is one of six programs in Cambodia's National Forest Program. He coordinates and advises with local Forestry Administration officers to develop the Forest Management Plan; coordinates with local Forestry Administration officers and community forestry management committees to develop the Community Forest Management Plan; and coordinates with key stakeholders to identify land use planning for the Community Forestry Management Plan.

KATHRYN SMITH-HANSSEN, PH.D. co-founded Community Forestry International in 1999 and is its administrative director. She has been involved with community forestry activities in Cambodia since 1998. She played an integral part in setting up the Forest Alliance for Cambodia in 2002 and oversaw the administration of the project office in Phnom Penh. She also helped edit many of CFI's extension projects in Cambodia, including print and video media, as well as CFI's publications on Cambodia. She was involved in CFI's early forest carbon studies in India, and helped co-design the early project design for Cambodia's first REDD+ project in Oddar Meanchey Province.

INDEX